BY THE EDITORS OF CONSUMER GUIDE®

ENERGY SAVERS CATALOG

G. P. PUTNAM'S SONS • New York

Table of Contents

Copyright© 1977 by Publications International, Ltd.
All rights reserved.
Published simultaneously in Canada by
Longman Canada Limited, Toronto.

Library of Congress Catalog Card Number: 77-78123

SBN: 399-12037-8 (cloth)
      399-12038-6 (paper)

This book may not be reproduced or quoted in whole or in part by mimeograph or any other printed means or for presentation on radio or television without written permission from:
Louis Weber, President
Publications International, Ltd.
3841 West Oakton Street
Skokie, Illinois 60076
Permission is never granted for commercial purposes.

Printed in the United States of America

Insulation

INADEQUATE INSULA-TION can permit an amazing amount of heated or cooled air to escape right through your walls and ceilings. To prevent this air from escaping—and thereby reduce your home's energy consumption and utility costs—you must wrap the entire living area of your house in the proper amount of properly installed insulating material.

What areas, specifically, should be insulated? Insulation should go between the floor and either the crawl space or unfinished basement, between the ceiling and the unfinished attic (between the attic and the roof if the attic is finished), and inside all exterior walls (including those adjacent to an unheated garage). The idea is to envelop the entire living area in insulating material.

Until recently, people referred to insulation requirements in terms of the number of inches of insulating material needed. Now, however, these rough quantitative figures have been replaced by a more realistic gauge—the R-value. A material's

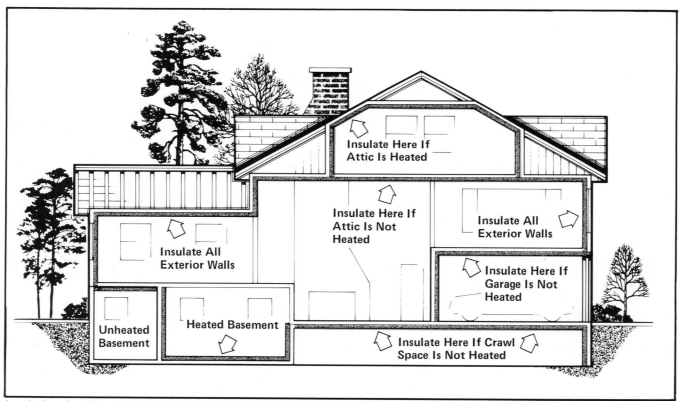

Insulation should go between the floor and either the unheated crawl space or basement, between the ceiling and unheated attic (between the attic and the roof if the attic is heated), and inside all exterior walls (including those adjacent to an unheated garage).

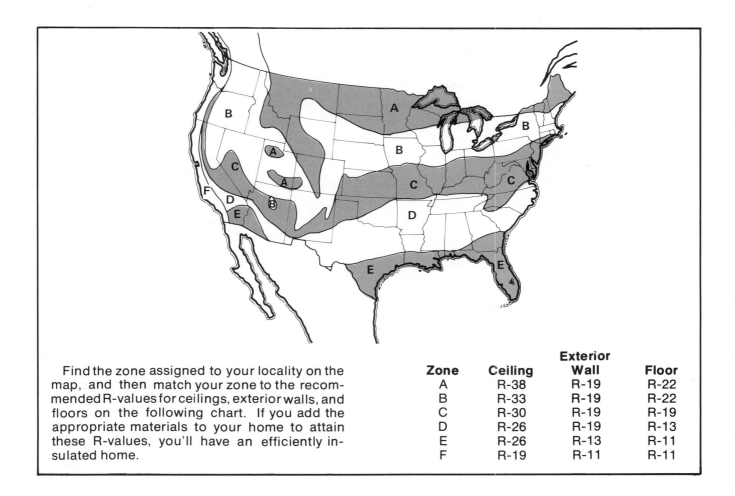

Find the zone assigned to your locality on the map, and then match your zone to the recommended R-values for ceilings, exterior walls, and floors on the following chart. If you add the appropriate materials to your home to attain these R-values, you'll have an efficiently insulated home.

Zone	Ceiling	Exterior Wall	Floor
A	R-38	R-19	R-22
B	R-33	R-19	R-22
C	R-30	R-19	R-19
D	R-26	R-19	R-13
E	R-26	R-13	R-11
F	R-19	R-11	R-11

R-value means its resistance to heat passing through it. The higher its R-value, the greater its insulating qualities.

The basic insulation materials generally available to the homeowner include a variety of R-values. But since these materials also vary in terms of installation ease, flammability, etc., selection of the right one for a particular application is not merely a matter of picking the one with the highest R-value. To make an intelligent choice, you should be familiar with all the properties of the most common insulating materials.

Types Of Insulation

HERE ARE some things to consider about the different types of materials commonly used for residential insulation.

Vermiculite. Due to its low cost, wide distribution, and easy installation, vermiculite is quite popular as a home insulating ma-

terial. A loose-fill material, vermiculite can be poured in and then raked, or it can be blown into place. It is an expanded mineral type of material that is easily introduced into hollow spaces, but it has the disadvantage of packing down from its own weight after several years. As it packs down, it loses thickness and thus its R-value is diminished. In addition, vermiculite absorbs moisture, causing even more packing, and the water itself greatly reduces the insulating qualities of the material. Therefore, vermiculite is not well suited to wall applications, where its settling tendency would leave uninsulated voids at the top of the wall space. Vermiculite is fire resistant.

Perlite. Perlite shares almost all the qualities outlined for vermiculite, but it offers a slightly higher R-value.

Fiberglass. A popular insulating material in wide dis-

Basic Material	Approximate R-value Per Inch Of Thickness*
Vermiculite	2.1
Perlite	2.7
Fiberglass	3.3
Rock Wool	3.3
Polystyrene	3.5
Cellulose	3.7
Urea Formaldehyde	4.5
Urethane	5.3

* Unfortunately, the per inch figure, when multiplied by thickness, is not a certain guide to discovering the material's R-value. Therefore, companies marketing insulation indicate the exact R-value on the product itself or on its package. Federal law also requires that bags of loose fill insulation show the R-value yield inch by inch.

tribution, fiberglass comes in batts, in rolls (blankets), and in pellets for loose-fill applications. It is relatively inexpensive and usually very easy to apply. Fiberglass itself is fire resistant, although the heavy paper facing frequently found on fiberglass batts and rolls is not fireproof. Fiberglass can also be faced with better materials that form a good vapor barrier, and it is available in an unfaced version for adding atop existing insulation. Its few disadvantages are that fiberglass is a skin irritant when handled and that it develops an odor when dampened.

Rock Wool. Rock wool offers almost the identical qualities described for fiberglass; even its cost and R-value are nearly the same. Like fiberglass, moreover, rock wool can irritate the skin when handled. About the only difference is the fact that rock wool does not develop an odor when wet.

Polystyrene. Popularly known as styrofoam, this type of rigid board insulation is an excellent material to use in new construction. Because it is combustible, however, polystyrene cannot be exposed in its finished state. It can be covered with wallboard, exterior siding, or other material as specified by local codes, and since it is moisture resistant, it can be used below grade; it is often used around slab foundations and has even been used as a base to provide extra insulation under a poured concrete slab. Polystyrene boards, very susceptible to gouges and dents, are generally attached to the studs of new construction during the framing.

Cellulose. This material has been much maligned as a fire hazard and is often referred to as "newspaper" insulation. When properly treated, however—as it is now required to be—cellulose is as fire resistant as fiberglass or rock wool. It offers a higher R-value than either fiberglass or rock wool, and it does not irritate the skin as those other types of insulation do. Cellulose comes in rolls (blankets), batts, or loose fill. In its loose-fill version, cellulose has a fine consistency, permitting blow-in installation through small access holes. Just make sure that the cellulose you purchase carries the brand name and treatment certification of a reputable manufacturer.

Urea Formaldehyde. A foam-in type insulation, urea formaldehyde can prove quite effective when installed properly. Its R-value is very high, and it offers excellent fire resistance. In addition, urea formaldehyde foam possesses excellent sound-absorbing qualities. Of course, a foam-in material has the advantage of completely filling any cavity in which it is injected, but it has the disadvantage of requiring installation equipment that is too expensive for the homeowner; thus, applying urea formaldehyde insulation is not a do-it-yourself project. Even some of the professionals in the insulation business lack the knowledge and experience to do a competent job. If improperly applied, the formaldehyde odor can linger. Therefore, if you elect to use this type of insulation, be sure to hire a qualified contractor to perform the installation.

Urethane. While urethane is the insulating material with the highest R-value, it shares the same disadvantages described for urea formaldehyde—plus the fact that it emits a noxious gas during a fire. On the other hand, the possibility of fume emission during a fire can be greatly reduced if the urethane is installed by a competent insulation professional.

Checking Your Insulation

YOU NEED to know the type and quantity of insulation already in your house before buying additional material. In most cases, you can check the attic insulation quite easily. With an unfloored attic, merely measure the thickness of the insulation with a ruler. Then, if you know what the material is, you can estimate its R-value. If you don't know the type of material, take a sample of it to an insulation dealer.

A floored attic presents a slightly different problem. If the boards are just butted together and nailed down, you can pry up a board to check. Usually the easiest place to start is at an exposed end at the attic entrance.

If the flooring is tongue and groove, you can drill a hole (a half-inch or larger) through the floor in an obscure corner. Be sure your hole is not over a joist and that you have a dowel of the proper size to plug the hole when you're finished checking the insulation. Use a flashlight to peer into the hole. If you can see that the insulation comes up to the flooring, you need only find out what type of material it is; the hook portion of a wire coathanger works well in retrieving a bit of the material.

If the insulation doesn't come up to the flooring, lower a probe into the hole until it touches the top of the insulation. Mark the probe at that point and then withdraw it. If you know the depth of the cavity and the thickness of the flooring, you can figure out how many inches of insulation are presently under your attic floor. If you don't know the depth of the cavity, however, you must push the probe through the insulation until it strikes the solid surface below. Mark the probe at that point and then compute the depth of the cavity by subtracting the thickness of the flooring from the total depth indicated on the probe.

To measure the insulation inside a wall, you must again find an opening. If possible, use existing openings such as those around electrical outlets. Before you start probing, however, turn off the electric current to the outlet selected.

Take the cover plate off the outlet and see if there is enough space to the side of the junction box to allow you to inspect the insulation with a flashlight. If not, widen the crack on the side opposite the side where the metal box is attached to the stud. Use a utility knife to widen a crack in sheetrock; use a cold chisel if the

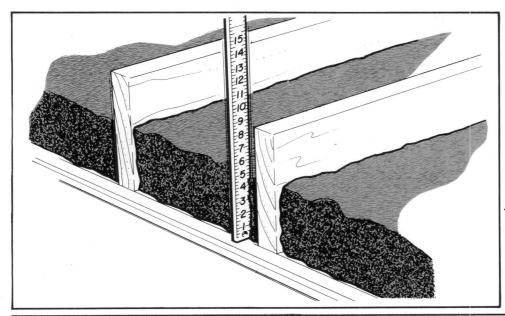

◀ With an unfinished attic floor, just measure the thickness of the insulation with a ruler. Then, if you know what the material is, you can estimate its R-value. If you don't know the type of material, take a sample of it to a reputable insulation dealer.

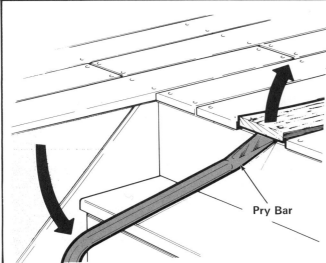

Pry Bar

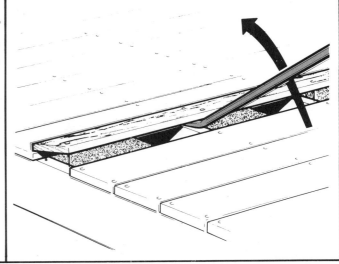

▲ If the boards of a floored attic are merely butted together, you can pry one up to check the insulation below. Start at an exposed end at the attic entrance (left) and work your way along as the floor board comes up (right).

If the boards of a floored attic ▶ are tongue and groove, drill a hole through an obscure portion (not over a joist), and use a flashlight to see whether the insulation comes up to the flooring. If it doesn't, lower a probe into the hole until it touches the material.

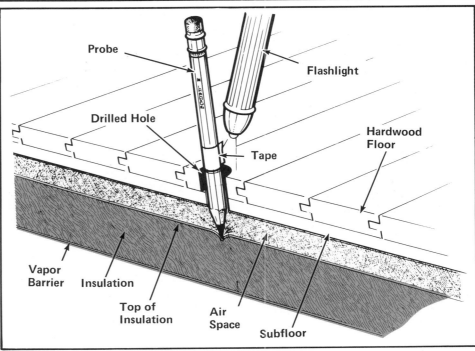

Probe

Flashlight

Drilled Hole

Tape

Hardwood Floor

Vapor Barrier

Insulation

Top of Insulation

Air Space

Subfloor

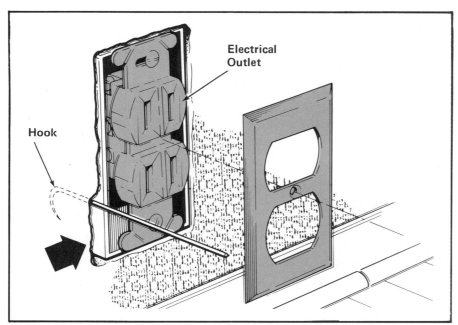

To check the type and amount of wall insulation, widen the opening around an electrical outlet on the side opposite where the metal junction box is attached to the stud. Pull out some of the insulation with a wire hook.

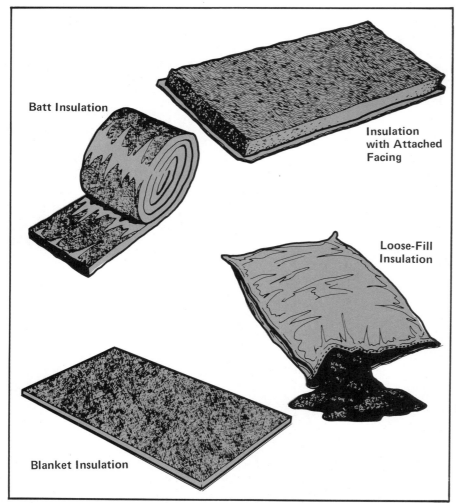

Insulation is available in batts, blankets, and loose fill. Batts and blankets can have an attached facing which serves as a vapor barrier, or they can come unfaced. Batts are easier to install in attics than blankets.

wall is plaster. Be careful, though, to widen only as much as the cover plate can hide. Then inspect with a flashlight and probe as you would a floored attic, pulling out a sample with a wire coathanger if you don't know the type of insulating material used.

Attic Insulation

NOW THAT YOU know what insulation you should have and the type and quantity of the material already there, you're ready to install the new material you need.

The easiest place to insulate in most homes is the attic. Fortunately, the attic is also the place where proper insulation has the most pronounced effect. In the winter most of the home's lost heat is lost through the attic, and in the summer the uninsulated attic acts as a heat collector, making the air conditioning system work harder than it should.

If you find that there is no insulation whatsoever in your attic, follow these steps for installing batts or blankets. You'll find that batts are generally easier to install than blankets in most attics.

Step 1. To determine how much insulation you need, measure the length and width of the attic and multiply the length by the width to arrive at the total square footage.

Step 2. Now, measure the distance between the joists. Most are on 16-inch centers (16 inches from the center of one to the center of the next), but some are on 24-inch centers. Buy batts or blankets of the correct width to fit between joists.

Step 3. For 16-inch centers, multiply the square footage by .90; that computation will give you the number of square feet of insulation required. With 24-inch centers multiply by .94.

Step 4. You must install a vapor barrier in attics having no insulation. The easiest way to lay down a vapor barrier is to install batts or blankets of insulation with a vapor barrier attached.

Step 5. Before you begin the installation, cut pieces of plywood to use as movable flooring, and carry up wide planks to serve as walkways. If you were to step onto the ceiling material, you would likely break right through it, but the joists will support your weight.

Step 6. If the attic is inadequately lighted, rig up a lighting system so that you can see what you're doing. A drop light suspended from a nail or hook will do the job.

Step 7. If you are installing fiberglass or rock wool, you must protect yourself with gloves, safety goggles, and a breathing mask. A hard hat is also a good idea to protect your head from protruding nails and low rafters.

Step 8. Now you're ready to lay down the insulating material. Start under the eaves and push the end of the blanket or batt in place with a long stick. Be sure to put the vapor barrier side on the bottom.

Step 9. Press the insulation down firmly between the joists.

Step 10. Continue until you reach the center of the room, and then work from the opposite end of the joists out to the center.

Step 11. When you must cut the insulating material—to fit around pipes and other obstructions—use a sharp knife with a serrated edge. In addition, you will find that the material is easier to cut when compressed with a scrap piece of board.

Step 12. Trim the insulation to fit around any vents, recessed lighting fixtures, exhaust fan motors, or any heat-producing equipment that protrudes into the attic. Allow three inches of clearance. Do not pull on any electrical wiring to move it out of the way.

Start under the attic eaves and push the end of the blanket or batt into place with a long stick. Be sure to put the vapor barrier side on the bottom and to press the insulation down firmly between the joists.

If you choose to insulate your uninsulated attic with loose fill, follow these steps.

Step 1. Install a vapor barrier.

Step 2. To calculate your total material needs, measure the square footage of the attic and then consult an insulation dealer. The dealer has a chart showing the maximum net coverage per bag at various thicknesses and the R-value for each thickness. The bags in which the loose fill is packaged also supply the same information.

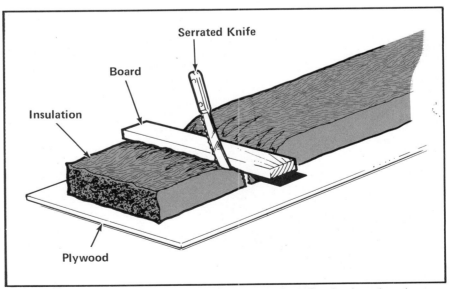

Use a sharp knife with a serrated edge to cut insulation. Cutting is easier when you compress the material against a scrap piece of board.

9

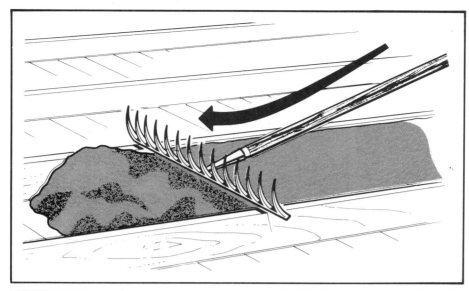

◀ *Pour loose fill into the spaces between the joists, and then spread and level the insulating material with a rake.*

To insulate a disappearing section of stairs, you must (A) construct a box over the stairs, (B) add a plywood door to the box, and then (C) staple insulation batts to all sides of the box and to the door.
▼

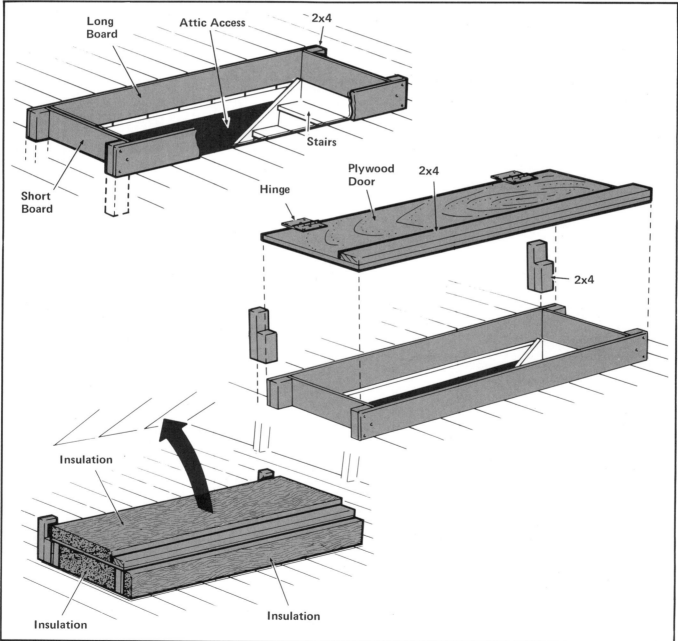

Long Board

Attic Access

2x4

Short Board

Stairs

Hinge

Plywood Door

2x4

2x4

Insulation

Insulation

Insulation

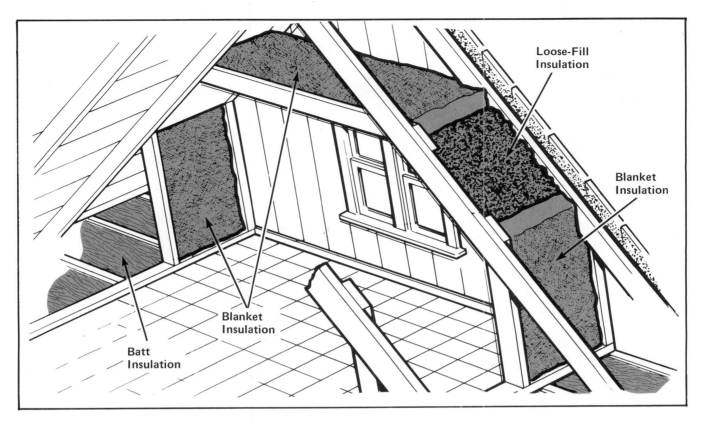

You can insulate a level span of attic ceiling with batts or blankets, stapling the material to auxiliary joists nailed to the rafters.

Step 3. You must be careful not to cover vents and heat-producing gadgets sticking up in the attic. Strips of insulating batts do a good job of guarding vents, while metal retainers made from tin cans will keep loose fill away from the other problems areas.

Step 4. Pour the insulation into the spaces between the joists, and then spread and level the material with a garden rake. If you want the joists to be covered with the loose fill, work from all sides back toward the attic access hole.

Step 5. Staple a batt of insulation material to the access cover.

You can use either batts or blankets or loose fill to add material to existing but inadequate insulation. Follow the same application procedure outlined for an uninsulated attic, but lay down unfaced batts or blankets instead of the versions with a vapor barrier attached. If you add loose fill, do not apply a vapor barrier over the existing insulation.

Special Attic Insulation Problems

IF YOUR access to the attic consists of a disappearing section of stairs, you must construct a box in the attic over the folded-up stairs. Add a plywood door to the top of the box, and then staple batts to all sides of the box and to the door.

Hand pack insulation around pipes and wires that come up through the floor of the attic, closing the holes around these elements.

If you decide to heat your attic (a relatively inexpensive way to provide extra living space), you should remove the existing insulation from between the joists before installing the finished attic floor. Remember that insulation should only go between heated and unheated areas, and keep in mind that the vapor barrier always faces the heated area. Staple insulation blankets between the rafters before you cover the rafters with paneling or sheetrock.

Insulate a level span of ceiling with batts or blankets stapled to auxiliary joists nailed to the rafters. In some cases, however, the blankets stapled between the rafters can be continued across this level area. Apply the ceiling finishing material after fastening the insulation to the framing.

Wall Insulation

THE NEXT MOST important places to insulate after the attic are the exterior walls. If you live in an older home in which you have access from the attic to the cavities between the wall studs, you can merely pour loose fill in the holes; just make sure that the cavities are closed at the bottom.

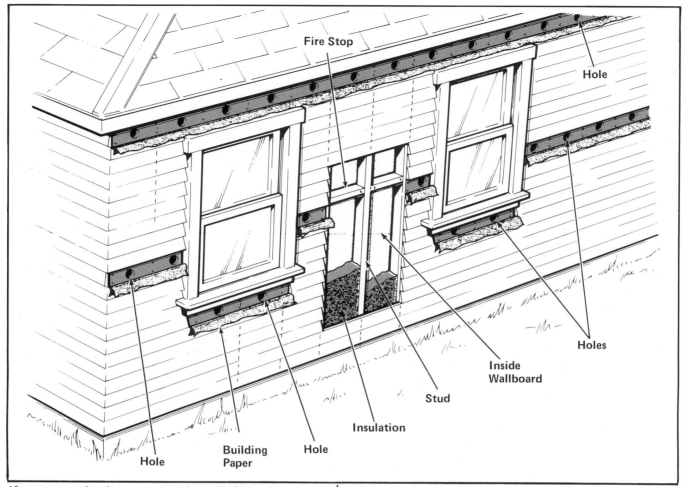

Fire Stop

Hole

Holes

Inside
Wallboard

Stud

Insulation

Hole

Building
Paper

Hole

Hole

If you cannot insulate your exterior walls from the attic, you'll have to cut holes in the walls about every 16 inches and blow the insulation in. If the walls have fire stops, you'll need a second row of holes just below the horizontal 2x4's. In addition, there must be holes beneath windows and other obstructions as well as at the top of each level. You can drill the holes from the inside or outside.

If you don't have this type of construction, you'll have to cut holes in the walls about every 16 inches and blow the insulation in. Additionally, if your exterior walls have fire stops (2x4's nailed horizontally between the studs), you'll need a second row of holes just below the fire stops.

Many exterior walls are built without fire stops, but you'll have to cut extra holes beneath windows and any other obstructions. A two-story house needs holes at the top of each floor in addition to all the other locations mentioned.

You can drill from the inside or from the outside. If the inside walls have decorative molding, you may be able to remove the molding, make holes behind it, and then cover the holes with the molding. If you have a brick home or one with metal siding,

then you certainly should drill from inside, but most other exteriors can be cut into and patched. Clapboard siding, for example, can often by pryed up without damage and replaced after the insulating is done.

To remove clapboard siding, insert a piece of sheet metal or a flat wide-bladed scraper under the piece of siding to be removed. The sheet metal or scraper will protect the siding from the pry bar that is wedged in over the protective metal and under the siding.

Pry gently until the siding comes up about a half inch. Then remove the pry bar and push the siding back in place; use a mallet if the siding won't go down. Pull out the nails you find sticking up, and then repeat the prying every 16 inches to remove all the nails. If the piece of siding does not

come out, look for a nail you missed or a paint seal holding the siding from above.

Wood shingles have hidden nails except at the top, although the top row may have some sort of overlap. If so, pry away the overlap. Now that the top row of shingles is exposed, you can usually pry them up without breaking any shingles. A flat spade slipped under the shingles provides good prying leverage. After the shingles are pried up slightly, the nail heads will be up far enough so that you can pull them out.

You may have to remove more than one row of shingles to find the right spots for the insulation access holes. For prying up shingles from the lower areas under obstructions, you might do well to invest in a specialized tool called a shingle nail remover. Asbestos shingles can also be removed with the shingle nail remover tool.

Removal of the facing material—whatever it may be—should reveal the building paper. Make a horizontal cut along the top of the paper, going all the way across the wall. Then make vertical cuts at each end, and fold the building paper down.

You are now ready to position the holes between each pair of studs. You should be able to spot the studs by noting the nails in the sheathing. Be sure to make the holes large enough to accommodate the nozzle of the insulation blower.

Before filling each cavity, make sure that the hole is open all the way down. Lower a weight on a string to check. If you encounter an obstruction, you'll need to drill another hole below.

Large tool rental companies, as well as large insulation dealers, rent insulation blower units. Be sure to get complete operating instructions.

When all cavities are filled with insulation, plug the holes with plastic inserts that snap in and lock. You can also nail tin can lids or sheet metal squares over the holes. Then staple the building paper back in place and replace the siding, using existing nail holes where possible.

When you must drill from inside, use a magnetic stud finder to locate the studs, and then drill or saw holes at the top of the wall big enough to accept the nozzle of the insulation blower. Drop a weight on a string to be sure the cavities are unobstructed, and then blow in the insulation according to the instructions supplied with your rental unit.

After the insulation has been blown in place, you must patch all the holes. Usually the holes are so large that they need backing to prevent the patching compound from falling into the cavity. If the loose-fill insulation does not provide enough backing, cram scraps from insulation batts into the holes, and then plaster over the openings with spackling compound. After the compound is dry, it must be painted.

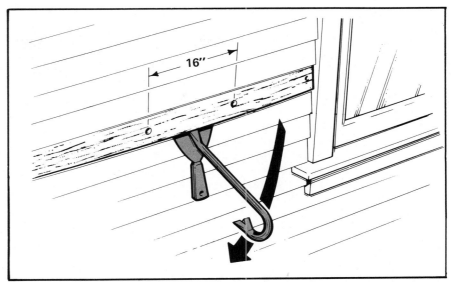

When removing clapboard siding, use a pry bar against a flat piece of metal to lift the siding up gently. Repeat the prying every 16 inches, pulling out the nails holding the siding as you go.

Floor Insulation

EXPOSED JOISTS in an unheated crawl space or basement allow you to provide a layer of insulation under the floor of your home. The best insulation for this type of installation are batts, since they are the easiest to handle. To install insulation batts between the floor joists, follow these steps.

Step 1. Estimate your material needs the same way you would if you were insulating the attic.

Step 2. After you buy the material, you have the problem of installing it so that it stays in place between the joists. Several options are open to you. Strips of wood lathe can be nailed across the joists about every 16 inches;

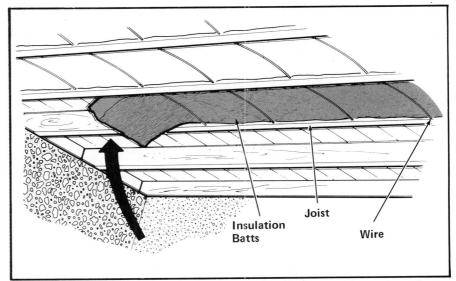

Insulation batts are easy to install between the exposed joists of an unheated crawl space or basement; heavy-gauge wire will hold the batts in place.

chicken wire strips can be stapled across the joists, leaving room between strips to work in the batts; heavy-gauge wires can be cut slightly longer than 16 inches and wedged between joists; or wire can be laced back and forth and held with nails. Whatever method you choose, though, must support the batts so that they don't sag. On the other hand, don't worry about an air space between the batts and the subflooring. If the batts are snug, the dead air space will actually contribute some R-value.

If you detect any problem with ground moisture, be sure to provide a ground vapor barrier. And if the moisture problem still exists after you install the vapor barrier—or if the crawl space is open—cover the bottom of the joists with low-grade plywood or staple a plastic vapor barrier in place.

Some homes have an insulated basement or insulation on the foundation to provide what is called a heated crawl space. The heated crawl space doesn't necessarily mean that heat is ac-

tually piped down there. The insulation around the foundation is designed to prevent heated air that reaches the crawl space from escaping.

Naturally, the basement or crawl space with insulation around the foundation does not provide nearly as effective a barrier to heat loss as does insulation under the floor. A heated basement, on the other hand, should be insulated around the foundation, while its ceiling (even if bare joists) should not. The warm air that rises from the

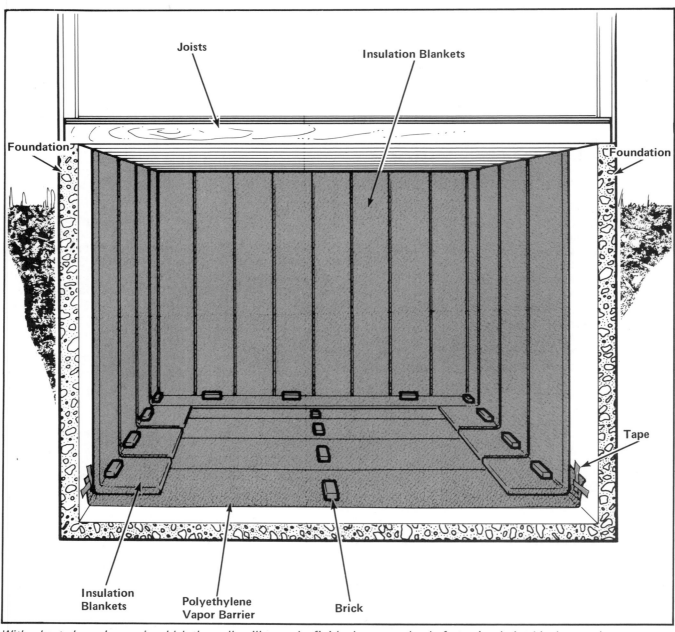

With a heated crawl space in which the walls will never be finished, you need only fasten insulating blankets at the top and pile bricks to hold the blankets against the walls at the floor.

heated basement will help warm the house.

Heated Basement And Crawl Space Insulation

IF THE HEATED basement is finished but uninsulated, handle it as you would any other room. Probably the most practical way to insulate it would be to blow in loose fill.

With an unfinished basement, you can choose from two practical alternatives: you can provide studs to which you staple or friction fit insulating blankets, or you can attach furring strips to which you apply rigid sheets of insulating material, covered over and made fire resistant. The latter method is easier in most instances, since the rigid insulation sheets can merely be glued to the wall with special mastics. Be sure to check your local building code to determine what types of wall covering are acceptable over this type of insulation.

With either method, though, be sure to cover the box joists and headers with blankets stapled to the subfloor above and the framing added below. The box joist is the joist that runs parallel to all other joists between the subfloor and the basement walls. The headers are the spaces between and at the end of all the other joists. Remember to install the insulation so that the vapor barrier side faces toward you.

With a heated crawl space in which the walls will never be finished, you need only fasten insulating blankets at the top and pile bricks to hold the blankets against the wall at the floor. Wider blankets can make the in-

stallation quicker, but a narrow crawl space is not going to be easy to insulate no matter what. Here are the steps to follow to do the job right.

Step 1. Start on a box joist side of the wall. Lay a plastic vapor barrier strip on the ground along that wall, leaving about two inches at each end to run up the wall. Use duct tape to secure the ends.

Step 2. Start at a corner and staple blankets to the box joist, trimming the blankets at ground level.

Step 3. Butt the next strip to the first and trim it, working all the way across.

Step 4. Trim the last panel to fit.

Step 5. Lay the next strip of vapor barrier, letting the two overlap by about six inches.

Step 6. Next, attach a blanket strip in the corner of the wall with headers. Trim to accommodate the joists, but don't trim at ground level; instead place a brick on the ground to hold the blanket. Then run the blanket out to the middle of the room.

Step 7. Go to the opposite corner, attach another strip in the

same manner, and run it out to meet the first strip. Trim both at the center and use bricks to hold each end down.

Step 8. Work your way on down each wall, overlapping vapor barrier strips as you go and trimming and attaching the blankets at the top. At floor level, however, extend the blankets toward the center no more than about two feet.

Step 9. When you reach the other box joist side, finish the wall and run corner blankets to the center.

Slab Foundation Insulation

A HOME built on a slab should have rigid insulation attached to the foundation all around its perimeter. Such insulation will make a significant difference in

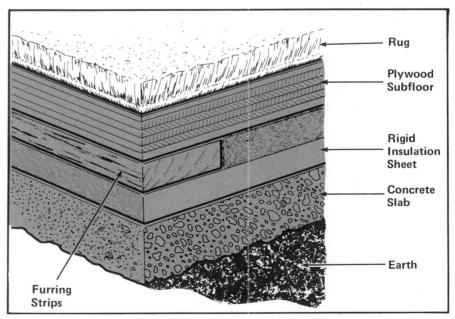

A home built on a slab should have rigid insulation attached to its foundation all around its perimeter. Apply the rigid sheets to the slab with mastic, attach furring strips and the plywood subfloor, and then install the finished flooring.

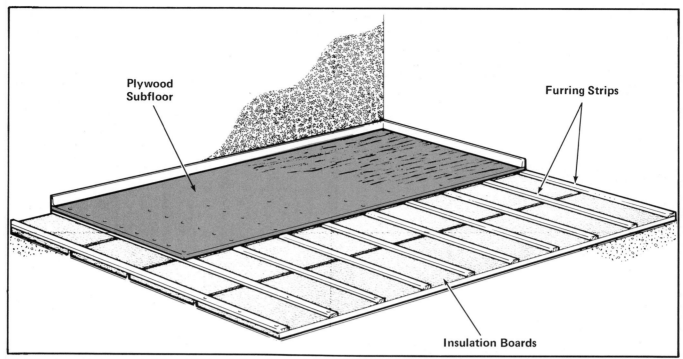

Plywood Subfloor

Furring Strips

Insulation Boards

Nail furring strips on top of the rigid insulation sheets, laying the furring strips on 16-inch centers to accommodate the subflooring. Naturally, if there is no vapor barrier under the slab, you should apply a vapor-repellent coating to the slab before attaching the rigid insulating sheets.

the temperature of the slab floor.

Since the insulating material should extend down below the frost line, the installation generally involves considerable digging. After the rigid insulating sheets are attached to the foundation with the specific mastic recommended by the insulation manufacturer, they should then be finished over with stucco, plaster, asbestos board, or whatever else is code approved.

When converting a garage or porch to living space, you can insulate the concrete floors quite effectively with rigid sheets. Just follow these steps.

Step 1. If there was no vapor barrier installed under the slab when it was poured, you should apply a vapor-repellent coating.

Step 2. Apply rigid sheets to the floor using the recommended mastic.

Step 3. Nail furring strips on top of the rigid insulation sheets, using masonry nails and a proper size hammer. Be sure to wear safety goggles. Lay the furring strips on 16-inch centers to accommodate the subflooring.

Step 4. Nail half-inch exterior grade plywood to the furring strips, and then apply your desired flooring.

Vapor Barriers

THE VAPOR BARRIER is not an actual energy saver, but it should always be included when insulation is installed for the first time. Since proper insulation increases the difference between the temperature on the inside and the temperature on the outside of a wall surface, it leads to increased condensation. Warm moist air passes through walls and ceilings and condenses when it reaches the colder surfaces. Moisture left on wall surfaces can cause paint failure,

mildew, and rotting, while moisture left inside the wall renders the insulating material almost worthless.

The key to stopping this condensation problem is the vapor barrier. It prevents warm moist air from passing through the walls or ceiling to the cold surfaces, and thus it prevents condensation from forming.

Adding a vapor barrier to an attic is simple when you are starting an insulation installation from scratch. You can buy batts or blankets with the vapor barrier facing attached, or you can put down a layer of special plastic material before pouring or blowing loose-fill insulation in place. Wide duct tape is useful in patching any tears or holes in the plastic (a very important procedure) and in taping two sections together.

A vapor barrier is very difficult to add to existing insulation or to wall insulation. If moisture is a problem, however, you can create an effective vapor barrier by applying to walls an oil-base enamel with a alkyd-based top coat. A penetrating floor sealer or even several coats of floor varnish can provide a satisfactory vapor barrier on existing floors. Even floor wax and wax coats on wood paneled walls help stop vapor-laden air from passing through.

There are other interior materials that act as vapor barriers. Vinyl wall coverings do a pretty fair job. Resiliant flooring also acts as a vapor barrier, but the tile squares are not as effective as sheet goods; the cracks between the tiles, small as they are, allow moisture to pass through. Although these materials can help when no proper vapor barriers can be installed, the conventional vapor barrier materials—four-mil polyethylene sheeting, aluminum foil and impregnated papers—still provide the best protection against the harmful effects of condensation.

Having It Done

EVEN IF YOU don't wish to do the insulating yourself, you can still save lots of money on heating and cooling by having the material installed by a professional insulation contractor. Here are the steps to follow in selecting a qualified contractor.

Step 1. Get three estimates from contractors you've heard do good work.

Step 2. Check the names of the contractors with the Better Business Bureau for recent references on residential work.

Step 3. Have a good idea of what work you want done before the contractors visit your home, but keep an open mind since your house may present unusual circumstances. Specify the R-value you want, and get the contractors to specify how much of what material will be used to attain that R-value. Any contractor who does not talk in R-value terms should be eliminated from consideration.

Step 4. Make sure that the contractor you choose carries insurance covering his employees as well as insurance covering damage to your home.

Step 5. Have a clear understanding of how the finished job is to look.

Step 6. While the work is being performed, check to make sure that all the bags or other material needed to do the job are of the type agreed upon and that they are actually installed in your home.

Insulation

Materials and Supplies

Thermography

There are now highly sophisticated infrared devices that can reveal in minute detail where heat is escaping from your home. The process is called thermography. The infrared devices can be used to pick out homes or plants with high heat losses, and they can be used to determine specific areas of individual homes that need additional insulation. A thermographic examination can be a money-saver if it reveals precisely those areas where insulation should be installed.

Thermographic equipment is far too expensive for the individual to buy, but a thermographic inspection performed by professional service personnel is quite reasonable. The operator does a scan of your house to determine where the trouble areas are. Most of these companies offer a follow-up scan after you have your home insulated to make certain that the heat loss problem has been eliminated.

Batts
And Blankets

If you need to add insulation to the existing insulation in your attic, consider adding unfaced (no vapor barrier) batts or blankets. Re Insul, made by Johns-Man-ville, is a fiberglass insulation made especially for such purposes. Installation involves merely laying the Re Insul on top of the existing insulation.

Fiberglass insulation batts from CertainTeed have a reflective thermal value in addition to their R-value. The reflective value is a product of the batt's aluminum foil facing. Since the vapor barrier should always rest directly against the attic floor, these batts should not be used as additions to existing attic insulation. They do make excellent new insulation, however. Available in 6 inch batts for use in the attic and 3½ inch batts for use in the walls, these foil-faced insulation batts are part of a whole

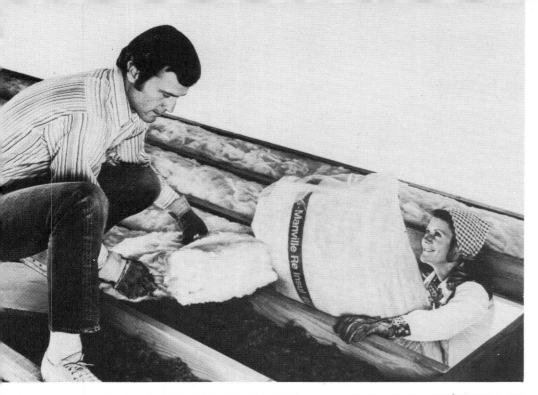

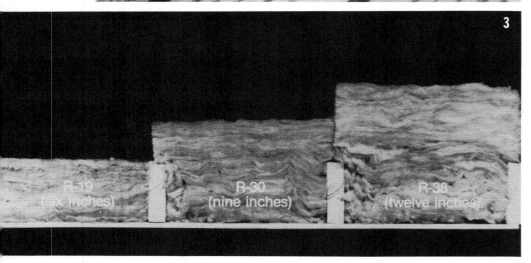

line of insulation products made by CertainTeed.(2)

Owens-Corning, a leader in fiberglass insulation, has developed two "high R" insulation blanket products designed to make it easier to achieve high R-values in less time. The blankets are available with or without a paper vapor barrier and are offered in thicknesses of R-19 (6 inches), R-30 (9 inches), and R-38 (12 inches). They are manufactured in 24-inch and 16-inch width.(3)

Owens-Corning recommends that the do-it-yourselfer opt for batts and blankets rather than loose fill. Blankets and batts yield a uniform thickness and density that means a consistent R-value, while loose fill is more difficult to install at a uniform thickness and tends to settle as time goes by. Batts and blankets are also easier to handle; they generally just lie in place between joists.(4)

One of the leading trade names in rock wool insulation is Premium Brand marketed by Rockwood Industries, Inc. Among the principal suppliers of Premium Brand at the retail level is Montgomery Ward. While the rock wool itself is fireproof, Premium Brand faced batts have a facing of asphalt-coated kraft paper that is flammable and cannot be left exposed after installation. Rock wool is also available as loose fill.

Heating ducts must be insulated, otherwise a great deal of heat will be lost before it reaches its destination. Standard Duct Wrap by CertainTeed is a permanent and fire-safe form of duct insulation that comes in rolls 4 feet wide and either 1½ inches or 2 inches thick. The rolls are 100 feet long for the 1½ inch wrap and 75 feet long for the 2 inch version. CertainTeed also makes duct liner material. (5)

Loose Fill

Installing loose-fill insulation by the blown-in method requires

(1) Johns-Manville Corp. fiberglass insulation. (2) Fiberglass batts by CertainTeed. (3) Owens-Corning fiberglass blankets.

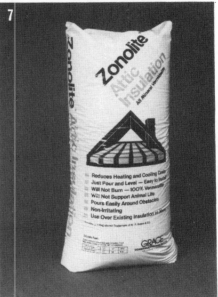

(4) Owens-Corning Fiberglas 6" Attic Insulation. (5) CertainTeed Duct Wrap. (6) Zonolite Handy Foam. (7) Zonolite Vermiculite Attic Insulation.

a special blower unit. Since this is expensive equipment, contact a large insulation retailer or tool rental agency to see whether you can rent one of the blowers for a short time. Doing it yourself, by the way, may not save you much money over having the loose-fill installed by professional insulation personnel, but many insulation contractors are running many weeks behind and if you need the job done quickly, you better plan to rent a blower unit.

Vermiculite is a popular loose-fill insulation material. Zonolite brand vermiculite from W.R. Grace and Company pours quickly into the voids and can be raked easily between joists. The Zonolite line of energy products also includes styrene foam in 2x8 foot sheets, Panelfoam, Handy Foam and Top-it as well as adhesives in tubes, cans and pails for applying these insulation products. (6)(7)

United States Mineral markets Cerama Fiber Insulation, a loose fill that can be poured or blown. This fireproof, inorganic and odorless ceramic mineral is similar to that used to protect the nose cone on spacecraft. Ce-

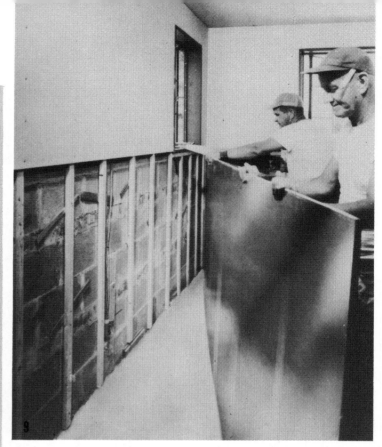

(8) Thermtron Wood Fiber Insulation. (9) U.S. Gypsum Panels. (10) United States Mineral Suprathane. (11) Swingline Stapler.

rama Fiber Insulation may be difficult to find at the retail level, but your insulation contractor should be able to obtain it for you.

Recent news stories have warned against the fire hazard posed by cellulose insulation. While cellulose is flammable before it is treated, all good cellulose material from reputable manufacturers is chemically treated with a fire retardant and poses no fire hazard or any other danger.

Thermtron insulation consists of recycled wood-based cellulosic fibers chemically treated with a fire retardant; it offers excellent R-values. Thermtron is available as loose fill to be blown in by Thermtron dealers.(8)

National Cellulose has created an insulating material of 100 percent wood-based cellulose fibers blended with fire-retardant and rodent/vermin-resisting chemicals. K-19 blow-in loose-fill insulation is a UL-listed Class A building material.

Rapco-Foam, one of the leaders in the foam-in field, has a dealer network that goes under the name of Homefoamers. Homefoamers are specialists in the application of urea formaldehyde insulation, which when properly installed can provide excellent results. The foam-in insulation is generally inserted into side walls of an existing home, with another type of insulation applied in the attic and under the floors.

Sheathing

InsulSheath's reflective vapor barrier skin is laminated to the urethane modified isocyanurate foam to increase its R-value. Due to its construction, InsulSheath replaces three steps in new or add-on construction: (1) sheathing, (2) vapor barrier, and (3) insulation. A product of Panel Era Manufacturing Ltd., InsulSheath should be used in combination with batts in a stud wall whenever possible.

If you are converting a garage or basement, you can install U.S. Gypsum's Foil-Back Sheetrock Gypsum Panels rather than adding a separate vapor barrier film. Installed like conventional sheetrock panels, the Foil-Back Gypsum Panels increase a home's R-value, preventing winter heat loss and summer heat gain. These panels are not recommended for use in high, hot, humid climates, however. (9)

Amspec's Totalwall Insulation System utilizes Styrofoam TG, which though treated with a fire-retardant agent will burn, and once ignited releases dense smoke. Installation must therefore conform to all local and national safety standards. Amspec's insulation/sheathing consists of 2x8 foot tongue-and-groove sheets designed to reduce air infiltration.

Suprathane rigid foam sheathing from United States Mineral consists of closed-cell urethane foam in 4x8 foot sheets. The sheets are sandwiched between aluminum foil skins reinforced with glass fibers. Available in ¾ inch, 1 inch, and 1¼ inch thick-

nesses, urathane sheathing can increase the insulation value of a wall.(10)

Cellofoam is an expanded polystyrene insulation from United States Mineral. Marketed in various widths, lengths and thicknesses, it can be used for perimeter insulation, cavity wall insulation, sheathing, and in various other ways. For example, Cellofoam is also offered in tapered form for use on flat roofs, where the taper provides a built-in drainage slope of ⅛ inch per foot.

Drew Foam from Drew International is another expanded polystyrene in sheathing form that is in general distribution. As with any of the popular rigid foam insulation materials, Drew Foam is flammable and must therefore be covered properly. Its 4.35 R-value per inch makes it a good addition within any exterior wall or as perimeter insulation for slab foundations. Drew Foam comes in 2x8 foot and 4x8 foot sheets, with special order lengths up to 16 feet. It is also available in loose fill for cavity walls, concrete block and attic insulation.

Siding

Considered strictly in terms of the R-value, aluminum, steel and vinyl siding contribute little to energy saving. From the standpoint of sealing leaks however, siding of any type can do wonders and if insulating material is added as backing for the siding, the R-value of walls can be increased significantly.

If you want to side or reside, investigate the possibility of having blown-in or foamed-in insulation put into the existing walls before the siding is installed.

Insulation Installation Tools

The Bostitch stapling hammer is often used to staple insulating

batts and blankets to studs or joints. It's a lightweight well-balanced tool that allows for very quick insulation applications.

Mar-Mac Insulation Supports come in very handy when applying batts or blankets under a floor between joists that are on 16 inch centers. These wire supports friction-fit between the joists to hold the insulating material securely in place; they can also be used for installing insulation between wall studs. Mar-Mac Insulation Supports are packaged 100 pieces to the package, and each pack will handle about 150 feet of insulation.

In order to ease the installation of mineral fiber or foam insulation on interior masonry surfaces, United States Gypsum has come up with Z-Furring Channels. The Z-shaped flanges attach to a masonry wall with standard masonry fasteners, and the strips hold the semi-rigid insulation in place. Sheet rock is then attached to the strips to form the outer wall. The Z-strips are designed to accommodate U.S. Gypsum's Thermafiber blankets or any rigid foam plastic in thicknesses of 1 inch, 1½ inches, 2 inches, or 3 inches.

In new construction or remodeling, there are often opportunities to add batts or blankets of insulation material between the stud walls, and stapling the facing to studs is the fastest way to do it. Installing a barrier also goes faster by stapling. Swingline makes a number of heavy-duty staplers for this purpose.(11)

Vapor Barrier

Polyethylene plastic, regularly used as a moisture barrier, is available in convenient rolls in thicknesses from two to six mils and in widths from three to twenty feet. Carry-Home Coverall, a product of Warp Brothers, is one popular and widely distributed brand. Get four mil or thicker plastic for vapor barrier purposes.

Weatherstripping

IF YOU HAD a six-inch square hole in the middle of your front door, you would certainly do something to plug it up. Yet, there are thousands of homes in which a crack an eighth of an inch wide exists all the way around the door, and this gap is just about the equivalent—in terms of air flow—of that six-inch square hole.

Gaps around doors and windows allow heat to escape during winter and cooled air to vanish during summer. Permitting these cracks to exist is, therefore, just like throwing dollars out the doors or windows. Fortunately, the simple procedure of installing weatherstripping can keep these dollars in your hands. Weatherstripping doors and windows in your home can reduce your heating/cooling bills by as much as thirty percent—to say nothing of the drop in drafts that can cause discomfort for you and members of your family.

Your home may or may not need weatherstripping. Luckily, there are some very simple ways to find out. Naturally, if you can feel cold air coming in around doors and windows on a windy day, you know the answer. If you are unsure, however, you can create your own concentrated windstorm at the precise spot where you suspect air might be leaking. Go outside with a hand-held hair dryer and have a helper inside move his or her hands around the door and/or window frame as you move the blower. When testing windows, of course, your helper can see where the blower is; but the two of you must establish voice signals when working on solid doors.

You may discover, when you finish testing, that all your doors and windows are airtight. Or you may find that a door or window that is airtight around three edges needs help along the fourth edge. What you will probably conclude, however, is that your home has several drafty areas and that you had better get busy installing weatherstripping in all of them.

Types Of Weatherstripping

THERE ARE several types of weatherstripping, with different situations calling for different types of material. All of the following types are available to the homeowner, and, unless otherwise stated, each type of weatherstripping can be used for either doors or windows.

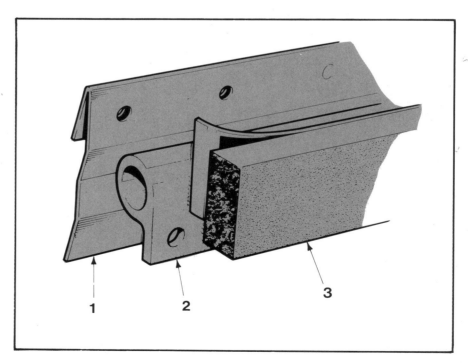

Three of the most popular types of weatherstripping include (1) spring metal, (2) tubular gasket, and (3) pressure-sensitive adhesive-backed foam. Each type can be used to weatherstrip both doors and windows.

Pressure-Sensitive Adhesive-Backed Foam. The easiest weatherstripping to apply and quite inexpensive, the pressure-sensitive adhesive-backed foam—available in both rubber and plastic—comes in rolls of varying lengths and thicknesses. When compressed by a door or window, the foam seals out the air. As an added advantage, these strips also provide a cushioning effect that silences slamming. Though not permanent, this type of weatherstripping can last from one to three years. Avoid getting paint on the material, however; paint causes the foam to lose its resiliency.

Spring Metal. These strips (V-shaped or single) are available in bronze, copper, stainless steel, and aluminum finishes. Most manufacturers package spring metal weatherstripping in rolls, and they include the brads that are needed for installation. Although this kind of weatherstripping seems like a simple installation, it requires a good deal of patience.

Self-Sticking Spring Metal. The peel-and-stick type of spring metal strips are like the standard spring metal strips just described, but they are far easier to install.

Felt. One of the old standbys and very economical, felt weatherstripping comes in a variety of widths, thicknesses, qualities, and colors (brown, gray, and black). Felt strips are usually nailed in place, but they are also available with a pressure-sensitive-adhesive backing.

Serrated Metal. This felt- or vinyl-backed weatherstripping offers the sturdiness of metal with the application ease of felt. Most manufacturers package their serrated metal weatherstripping in rolls and include brads for installation.

Tubular Gasket. Made of extremely flexible vinyl, tubular gasket weatherstripping is usually applied outside where it easily conforms to uneven places. Available in white and gray, it cannot be painted because paint

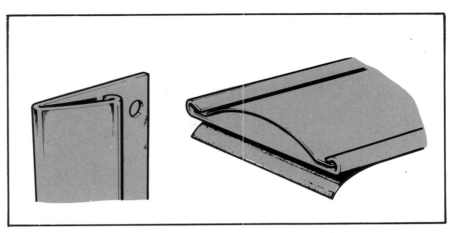

Spring metal strips are available in V-shaped configurations (left) and with an adhesive backing (right) for peel-and-stick applications.

Felt weatherstripping (left) has been superseded by other types, but still can be useful. Serrated metal (right) is more sturdy than plain felt.

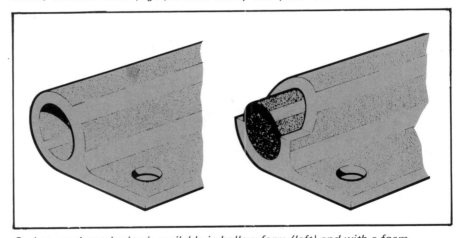

Gasket weatherstripping is available in hollow form (left) and with a foam filling (right). The foam-filled gasket is stronger but less flexible.

causes the tube to stiffen and lose its flexibility.

Foam-Filled Tubular Gasket. Adding foam to the tubular part of the gasket just described gives it extra insulating qualities and extra strength. Moreover, the foam-filled tubular gasket will hold its shape better than the hollow tube type, and like the hollow tube, it should not be painted.

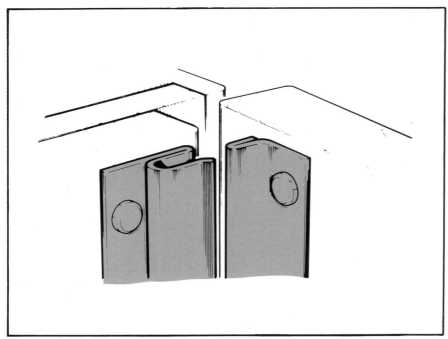

Interlocking metal weatherstripping can provide a secure seal as long as the separate pieces fit together as they should. Installation is tricky, and maintenance requires careful examination for bent pieces.

in fact, do damage to surrounding surfaces. Therefore, no step-by-step installation instructions are provided for this type. If you already have interlocking metal weatherstripping installed, keep it working right by straightening any bent pieces.

Casement Window Gasket. These specially made vinyl gaskets slip over the lip of the casement frame. No adhesives or tools—except scissors for cutting the gasket to the proper length—are needed. This weatherstripping is generally available only in shades of gray.

Jalousie Gaskets. Jalousie gaskets are clear vinyl tracks which can be cut to fit over the edges of jalousie louvers. They snap in place for a friction fit.

Door Sweeps. When there is a gap at the threshold under a door, attach a door sweep to create a tight seal. Door sweeps come in wood and felt, wood and foam, metal and vinyl, and in a spring-operated version that is mounted on the outside of a door which opens inward.

Interlocking Metal. This type of weatherstripping requires two separate pieces along each edge because one part fits inside the other to form the seal. One piece goes on the door while the other is attached to the jamb. Since installation generally requires cutting (rabbeting), interlocking metal weatherstripping may well be beyond the capabilities of the average homeowner. In addition, should the pieces get bent, they would no longer seal and could,

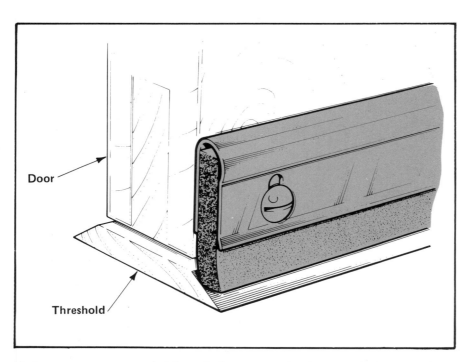

Door

Threshold

A door sweep can create a tight seal when a gap exists between the bottom of the door and the threshold. Door sweeps can be made of wood and felt, wood and foam, or metal and vinyl. All are effective in sealing out drafts.

Weatherstripping Windows

Wooden Windows

DOUBLE-HUNG wood windows almost always require weatherstripping, although if the top sash is never opened, you can solve an air leak problem by caulking to seal any cracks. You may find it advantageous to use more than one type of weatherstripping to complete the job. Be sure to follow the correct installation procedures for each type.

Spring metal weatherstripping

fits into the tracks around the windows. Each strip should be about two inches longer than each sash so that the end of the strip is exposed when the windows are closed. Position the vertical strips so that the flared flange faces toward the outside. The center strip should be mounted to the upper sash with the flare aimed down, while the other horizontal strips are mounted to the top of the upper sash and the bottom of the lower sash with the flared flange facing out. Cut the spring metal weatherstripping to allow for the window pulley mechanisms.

After you measure and cut the spring metal strips to size (use tin snips), follow these instructions to attach the strips to the window frame.

Double-hung wood windows (left) almost always require weatherstripping that completely surrounds both sashes. Flare the edge of a spring metal strip (right) with a screwdriver to render a snug fit.

Step 1. Remember to position the flange on either the V strip or the single strip so that the flared edge faces outside.

Step 2. Position the strip properly and note any hinges, locks, or other hardware that might interfere; trim away the metal where needed. Then trim the ends of the strip at an angle (miter) where vertical and horizontal strips meet.

Step 3. Tap in one nail at the top and one nail at the bottom of the strip. Do not put in any more, and do not drive the top and bottom nails all the way in. Since some V strips do not come with nail holes, you may have to make pilot holes with an ice pick or awl.

Step 4. Check to make sure that the strips are straight and properly positioned.

Step 5. Drive a nail in the center of the strip, but again only part way. Then add nails in between. To avoid damaging the strip, never drive any of the nails all the way in with the hammer. Instead, drive the nails flush with a nail-set.

Step 6. Flare out the edge of the strip with a screwdriver to render a snug fit.

Pressure-sensitive types of weatherstripping can be used only on the friction-free parts of a wooden window—e.g., the lower sash or the top of the upper sash. If the strip were installed snugly

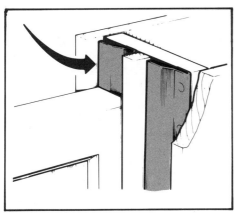

Position the flange on spring metal (above) so that the flared edge faces outside. Apply pressure-sensitive types (right) only on the friction-free parts of a wooden window.

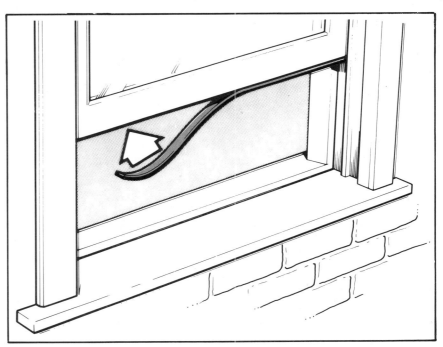

against the gap between upper and lower sashes, the movement of the window would pull it loose.

To install pressure-sensitive foam weatherstripping, first select (if possible) a warm day to do the work, and then proceed according to the following instructions. The adhesive forms a better bond when applied on a day in which the temperature is at least 60 degrees Fahrenheit.

Step 1. Clean the entire surface to which the weatherstripping is to be attached. Use a detergent solution, and make certain that no dirt or grease remains. If pressure-sensitive weatherstripping had been installed previously, use a solvent to remove any old adhesive.

Step 2. Dry the surface.

Step 3. Use scissors to cut the strip to fit, but do not remove the backing paper yet.

Step 4. Start at one end and slowly peel the paper backing as you push the sticky foam strips into place. If the backing proves stubborn at the beginning, stretch the foam until the seal between the backing and the foam breaks.

To install self-sticking spring metal weatherstripping on wooden windows, follow these instructions.

Step 1. Measure and cut the strips to fit.

Step 2. Clean the surface where the strips are to go.

Step 3. Put the strips in place without removing the backing paper, and mark the spots for trimming (e.g., hardware points and where vertical and horizontal strips meet).

Step 4. Peel off the backing at one end and press the strip in place, peeling and pressing as you work toward the other end.

Felt strips are somewhat unsightly for sealing gaps on wooden windows. There are places where felt can be used to good advantage, however. Attach felt strips to the bottom of the lower sash, the top of the upper sash,

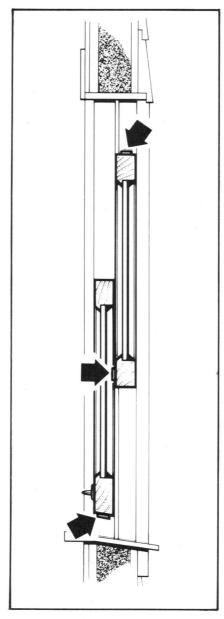

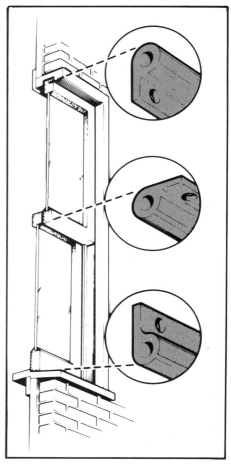

Although somewhat unsightly on wooden windows, felt strips (left) can be placed on the bottom of the lower sash, top of the upper sash, and to the interior side of the upper sash. Tubular gaskets (above) are also unsightly and should be installed only on window exteriors.

and to the interior side of the upper sash. The strips will then function as horizontal gaskets.

To weatherstrip with felt strips, follow this procedure.

Step 1. Measure the felt and cut the strip from the roll with scissors. Keep in mind that felt strips can go around corners.

Step 2. Push the material snugly against the gap.

Step 3. Nail the ends of each strip first, but do not drive these nails flush; leave room to pry them out.

Step 4. Start at one end and drive a tack every two to three inches,

pulling the felt tight as you go. If you find slack when you reach the other end, remove the nail, pull to tighten, and trim off any excess.

To install pressure-sensitive felt, follow the same steps as you would to attach pressure-sensitive foam.

Tubular types of weatherstripping are also unsightly. They can be used, however, when installed on the outside of the window. If the window is easily accessible from outside the house, therefore, tubular weatherstripping is worth considering; it can also be used to improve existing weatherstripping.

Install tubular and foam-filled gaskets in the following manner.

Step 1. Measure the strips, and then cut them to size with scissors. Cutting all the strips for a given window at one time will save you trips up and down the ladder later on.

Step 2. Position each strip carefully, and drive a nail into one end.

Step 3. Space nails every two to three inches, pulling the weatherstripping tight before you drive each nail.

Metal Windows

MOST METAL windows are grooved around the edges so that the metal flanges will interlock and preclude the need for weatherstripping. Sometimes, though, gaps do exist, and you must apply weatherstripping in such instances.

Generally, the only kind of weatherstripping that can be applied to metal windows is the pressure-sensitive type. Screws would go through the metal and impede movement of the window.

Apply the weatherstripping to the top of the upper sash (if it is movable) and to the bottom of the lower sash; these are usually the only spots where metal windows allow for air movement. If you find other gaps, however, attach a vinyl tubular gasket to the area with a special adhesive formulated to hold vinyl to metal.

Sliding Windows

SLIDING WINDOWS, those in which the sash moves laterally, come in both wood and metal frames. Weatherstrip the wooden frames much as you would a double-hung window turned sideways. If only one sash moves, weatherstrip it and caulk the stationary sash. For metal frames, follow the instructions for weatherstripping standard metal windows.

Jalousies And Casements

SPECIAL GASKETS are available which are designed especially for

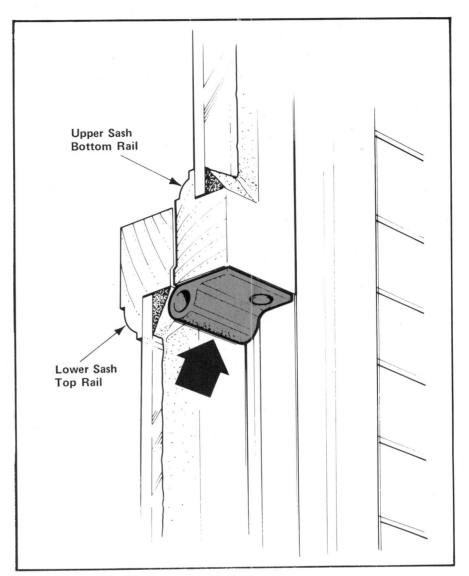

Upper Sash Bottom Rail

Lower Sash Top Rail

Nail strips of tubular gasket weatherstripping to the outside of the window in places where they would not be visible from inside.

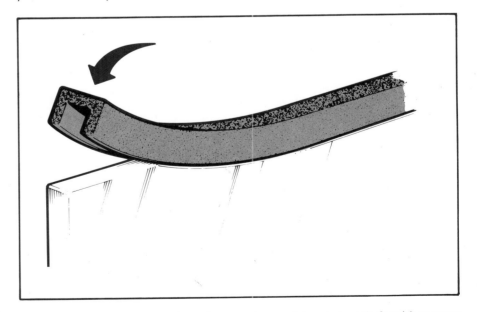

To weatherstrip jalousie windows, just cut the special gasket made for this purpose to size, and snap it into place along the edge of the glass.

27

sealing gaps in jalousie and casement windows. To weatherstrip jalousies, measure the edge of the glass louver, cut the gasket to size with scissors, and snap the gasket in place. To weatherstrip casement windows, measure the edges of the frame, cut strips of gasket to size, miter the ends of the gasket strips where they will intersect, and slip the strips in place over the lip of the frame.

Weatherstripping Doors

ALL FOUR edges around doors can permit air to leak in and out of your house. In fact, the average door has more gaps than does a loose-fitting window. Doors, moreover, don't run in grooves as windows do, and any crack area around a door is likely to be far greater than the analogous area around a window.

Before you start weatherstripping, however, inspect your door to be sure it fits properly in the frame opening. Close the door and observe it from the inside. Look to see that the distance between the door and the frame is uniform all along both sides and at the top. The distance does not have to be precisely the same all the way around, but if the door rests crooked in the frame, the addition of weatherstripping may make it impossible to open or close. Naturally, if there is great variance in the opening between the door and frame, it will be difficult to fit weatherstripping snuggly at all points; gaps may well be the result.

The cause of most door problems can usually be found in the hinges. The first thing to do, therefore, is open the door and tighten all the hinge screws. Even slightly loose screws can

cause the door to sag. If the screw holes have been reamed out and are now too big to hold the screws, you can use larger screws as long as they will still fit in the hinge's countersunk holes. If even the larger screws won't work, pack the holes with toothpicks dipped in glue, and use a knife to cut off the toothpicks even with the surface. Now the screws have new wood in which to bite.

Sometimes the door must be planed off to prevent binding. If so, you can usually plane the top with the door still in place. Always move the plane toward the center of the door to avoid splintering off edges. If you must take wood off the sides, plane the hinge side and always move toward the edges.

Spring metal is quite popular for door weatherstripping. It works effectively when installed properly, and it is not visible with the door closed. Most manufacturers include the triangular piece that fits next to the striker plate on the jamb in the packages designated as door kits. If none is provided (or if you choose to buy spring metal in bulk), you can purchase the triangular piece separately.

After you measure and cut the spring metal strips to size, follow these instructions to attach the strips to the door stop.

Step 1. Position the side strips so that the flared flange almost touches the stop.

Step 2. Trim away the metal where needed to accommodate any hinges, locks, or other hardware.

Step 3. Tap in one nail at the top and one nail at the bottom of each side strip. Do not put in any more nails, and don't drive the top and bottom nails all the way in. If the strips do not have prepunched holes, make pilot holes with an ice pick or awl.

Step 4. Check to make sure that the side strips are straight and properly positioned.

Step 5. Drive a nail in the center of a side strip, but again only part way in. Then add the nails in between. To avoid damaging the strip, never drive any of the nails all the way in with the hammer. Instead, drive the nails flush with a nailset. Repeat the procedure for the other side strip.

Step 6. Put the top strip in last and miter it to fit.

Step 7. Flare out the edge of each strip with a screwdriver to render a snug fit.

Self-sticking spring metal can be used in the same places. To install self-sticking spring metal weatherstripping around doors, follow these instructions.

Step 1. Measure and cut the strips to fit.

Step 2. Clean the surface where strips are to go.

The triangular piece of spring metal fits next to the striker plate.

Step 3. Put the strips in place without removing the backing paper, and mark the spots for trimming (hardware points and where vertical and horizontal strips meet).

Step 4. Peel off the backing at one end and press the strip in place, peeling and pressing as you work toward the other end.

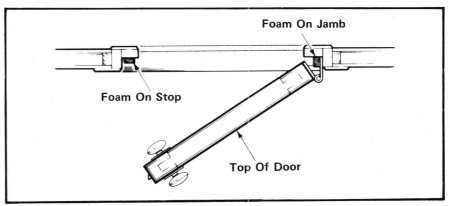

Attach strips of pressure-sensitive foam on the hinge side of the door jamb and on the door stop. The adhesive backing on the foam will form a secure bond only when applied to a clean, grease-free surface.

Pressure-sensitive foam types of weatherstripping can be easily installed around most doors. The foams are effective, but they have a shorter life span than other weatherstripping materials.

To install pressure-sensitive foam weatherstripping, first select (if possible) a warm day to do the work and then proceed according to the following instructions. The adhesive forms a better bond when applied on a day in which the temperature is at least 60 degrees Fahrenheit.

Step 1. Clean the entire surface to which the weatherstripping is to be attached. Use a detergent solution, and make certain that no dirt or grease remains. If pressure-sensitive weatherstripping had been installed previously, use a solvent to remove any old adhesive.

Step 2. Dry the surface.

Step 3. Use scissors to cut the strip to fit, but don't remove the backing paper yet.

Step 4. Start at one end and slowly peel the paper backing as you push the sticky foam strips into place. If the backing proves stubborn at the beginning, stretch the foam until the seal between the backing and foam breaks.

Step 5. Attach the strips on the hinge side to the door jamb.

Step 6. Attach the other two strips to the door stop.

Step 7. If the corner of the door catches the weatherstripping as you close it, trim the top piece of foam at the hinge side.

Serrated metal weatherstripping, usually with a felt strip insert running the length of the serrated groove, also can be used to seal air gaps around doors. To install this type of weatherstripping material, follow these steps.

Step 1. Measure the length of strips required, and then use tin snips or heavy-duty scissors to cut the serrated metal material to the proper lengths.

Step 2. Nail each strip at both ends.

Step 3. Add a nail to the center of each strip.

Step 4. Drive the remaining nails, spacing them every two to three inches.

Sweeps And Thresholds

YOU MUST treat the gap at the bottom of the door differently than you do those on the sides and along the top.

The hump on the floor along the bottom of the door is called the threshold. A threshold can be made of either wood or metal, with many of the metal types featuring a flexible vinyl insert that creates a tight seal when the door closes against it. Other thresholds consist of two parts —one on the floor and a mating piece on the bottom of the door —that interlock to form a weathertight barrier.

In most cases, the threshold with a flexible vinyl insert—since it adapts itself easily to most door bottoms—is the best one for a novice weatherstripper to install.

Interlock systems are quite effective when properly installed, but they require a perfect fit or else they will not work satisfactorily.

Wooden thresholds often wear down to the point where they must be replaced. This is an easy installation, and there are many types of replacement thresholds

Many metal thresholds have vinyl inserts that create a tight seal against the bottom of the door.

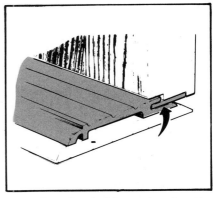

Two-piece thresholds consist of mating parts (one on the floor, one on the door) that interlock to form a tight seal.

from which to choose. Most are aluminum and come in standard door widths, but if your door is not standard width, trim the aluminum threshold with a hacksaw.

To install a replacement threshold, follow these instructions.

Step 1. Remove the old threshold. If it is wood, there are two ways to remove it. In most cases, you can pry it up after removing the door stops with a small flat pry bar or putty knife, but you must work carefully and slowly.

Step 2. In some cases, the jamb itself rests on the threshold. If this is what you find, then it is best to saw through the old threshold at each end. Use a backsaw placed right against the jamb, and saw down through the threshold, being careful not to scar the floor. Once you make the cuts, the threshold should be easy to pry up. But if prying fails to do the job, use a chisel and hammer to split the piece.

Step 3. Metal thresholds are frequently held down by screws concealed under the vinyl inserts. Remove the screws and the threshold will come up easily.

Step 4. Install the replacement

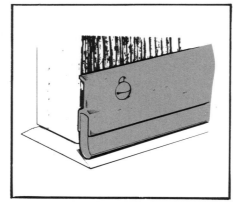

Most door sweeps (left) are attached to the inside of the door with nails or screws. Another type (right) is attached to the outside and flips up to pass over the threshold when the door is opened. It flips down when the door is closed to provide a snug seal against the threshold.

threshold by driving screws through the metal unit and into the floor. If the idea of an aluminum threshold offends you, you can cut a replacement from wood, using the original one as a pattern. Then install a door sweep to seal the gap.

If, for some reason, a gap exists between the door and the threshold, you can generally solve the problem by adding a door sweep. Most sweeps are attached with nails or screws to the inside of the door. Just cut the sweep to size and close the door.

Then tack both ends of the sweep to the door, after which you can install the remaining nails or screws. If you are using screws, drill pilot holes first.

Some types of sweeps slip under the door and wrap around the bottom. Still another type fits on the outside, with a section of it flipping upward to miss the threshold when the door is opened. When the door is closed, this section flips back down to provide a seal against the threshold. You can adjust this flip-style door sweep so that it renders a snug fit.

Weatherstripping

Materials and Supplies

Pressure-Sensitive Weatherstripping

The Weatherseal Group of the Schlegel Corporation has introduced a new pressure-sensitive weatherstripping called Polyflex. Just peel off the backing and stick the plastic weatherstripping piece in place on metal or wood. As with all pres-

sure-sensitive materials, Polyflex sticks best to a clean surface. It can be cut easily with a pair of household scissors. (1)(2)(3)

A self-sticking foam type weatherstripping (e.g., open cell polyurethane, closed cell vinyl, neoprene sponge, and sponge rubber) comes in several lengths, widths, and thicknesses. Leading makers of foam type weath-

erstripping include Macklanburg-Duncan, Mortell, W. J. Dennis, Deflect-o, Pemko, and Teledyne Mono-Thane; little difference in quality exists among the major brands. While this type of weatherstripping must be considered temporary because it generally does not last much longer than a year, it is so easy to apply that it has become one of the great success stories in the weatherstripping field. The se-

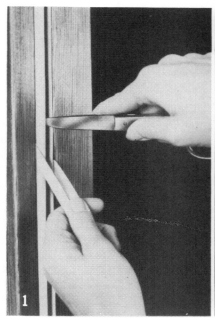
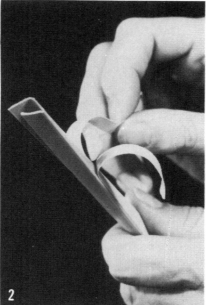
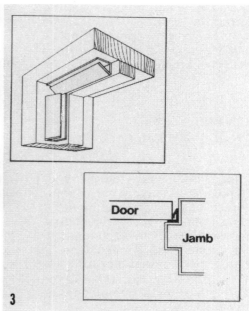

(1), (2) & (3) Polyflex weatherstrip installation. (4) Frost-King sponge rubber weatherstrip tape. (5) Teledyne Mono-Thane Foamedge weatherstrip. (6) Dennis Self-Seal cushion weatherstrip. (7) Mortell vinyl-foam tape. (8) Deflect-o Energy Seal foam weatherstrip.

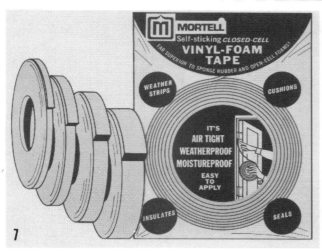
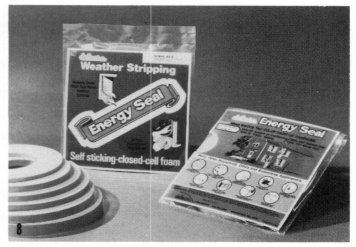

cret of a secure application is to clean the surface thoroughly before attempting to put the foam strips in place. If the strips should later come loose in spots, they can be reglued with a contact cement. (4)(5)(6)

Open cell type foam weatherstripping must be compressed by a door or window to be effective. If the situation dictates that the foam will be compressed only slightly, use a closed cell foam instead. Makers of self-sticking closed cell vinyl foam include Mortell, W. J. Dennis, Thermwell, Deflect-o, and several others. (7)(8)

While it is not a true weatherstrip, Mortell's transparent plastic tape can serve the same purpose when applied around windows from the inside. Made of a heavy-duty poly material, it adheres to any clean, dry surface and is scarcely visible. (10)

Gasket Weatherstripping

Inner-Seal, made by the W.J. Dennis & Company, is a reinforced gasket type weatherstripping that has spring steel wire molded into its live sponge rubber core. The entire piece is coated with neoprene. As a result, Inner-Seal offers the strength of metal combined with the flexibility of sponge rubber. It is installed with small nails. (9)(12)

Vina-Foam is the brand name for a gasket made of tough vinyl and filled with polyurethane foam from Macklanburg-Duncan. The foam stays resilient even in extremely cold weather, and the vinyl covering is unharmed by sunlight, paint, salt air, or other corrosive forces. Vina-Foam is generally tacked in place (M-D furnishes tacks in the packet), although it can also be installed with a staple gun. Foam-filled gaskets hold their shape better than do the hollow gaskets. (11)

By utilizing the same principle found on refrigerator doors, Portaseal offers a full doorstop with a magnetic core sealing gasket for use on metal-clad exterior doors. Magnet-Stop has a self-adjusting gasket between its wooden stop (which is nailed to the door jamb) and its magnetic core. It holds the closed door in a vacuum-like grip along the header and latch side—forming a total airtight seal along these two sides—and yet it releases the door easily. The hinge side must be sealed with a standard Portaseal vinyl stop. (14)

Even though the weatherstripping qualities of gasket type products depend, for the most part, on the way the tube fits against the surface, the lip used for nailing or stapling can be a weak spot. If the material rips here—and it often does—the gasket may no longer stay in place. Therefore, the most popular type of gasket weatherstripping features a serrated aluminum strip as reinforcement. The serrated metal provides strength without sacrificing all the flexibility of the vinyl gasket. Pemko is one of the major manufacturers of this type of reinforced weatherstripping. (13)

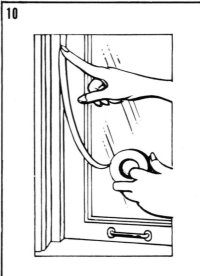

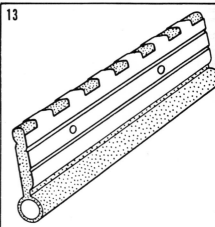

(9)&(12) Inner-Seal weatherstripping. (10) Mortell transparent plastic tape. (11) Vina-Foam gasket installation. (13) Pemko reinforced weatherstrip. (14) Magnet-Stop doorstop installation. (15) H-4 Foldback weatherstrip. (16) H-6 Foldback weatherstrip. (17) Pemko metal weatherstrip. (18) Frost-King metal weatherstrip.

Spring Metal Weatherstripping

A difference in shape plus a pressure-sensitive backing make Kel-Eez spring metal weatherstripping unique. Packaged for either door or window applications with just the right amount of material for each, Kel-Eez door kits are also available with a slip-on door bottom seal that shares the weatherstripping's dome shape. The door bottom is held in place on both sides of the door by its own spring tension.

Macklanburg-Duncan markets two variations of the basic spring metal type of weatherstripping. H-4 Foldback is designed to be nailed in place, while H-6 Foldback comes with an adhesive backing which needs only to be pressed in place. The advantage of foldback type over the single strip is that it tends to space itself better against the surface it is protecting. The single strip, in contrast, often must be pried out with a screwdriver in order to seal the gap properly. Remember that you must start with a clean dry surface when installing the pressure-sensitive type, and choose a day when the temperature is at least 60 degrees Fahrenheit; at colder temperatures the tackiness of the adhesive is retarded. This caution applies only to the installation of the weatherstripping, however. The cold winter weather in no way affects the adhesion of pressure-sensitive weatherstripping properly installed. (15)(16)

Typical of many good spring metal weatherstripping available are those that come under the Pemko and Frost-King brand names. Both types are packaged with the necessary tacks for installation. There are many manufacturers of this type of weatherstripping, and you can choose from among bronze, tempered aluminum, copper, and stainless steel strips. Little difference exists among the various brands and metals in terms of effectiveness, although strips with pre-drilled holes are much easier to install. For applications around doors, look for packets that include the piece that seals the gap around the latch. (17)(18)

Other Forms Of Weatherstripping

While there are better weatherstripping materials, the old standby—felt—can still do a

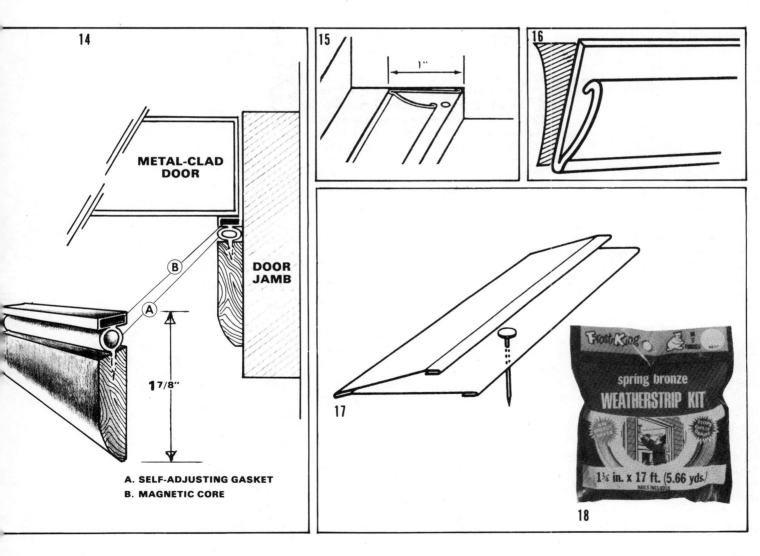

METAL-CLAD DOOR

DOOR JAMB

B

A

1 7/8"

A. SELF-ADJUSTING GASKET
B. MAGNETIC CORE

14

15

1"

16

17

Frost King

spring bronze
WEATHERSTRIP KIT

1¼ in. x 17 ft. (5.66 yds.)
NAILS INCLUDED

18

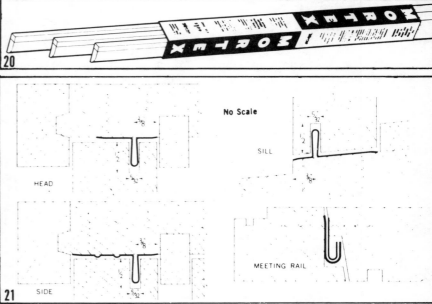

good job of sealing gaps around doors, windows, and air conditioners. Several companies manufacture felt weatherstripping, and you should have no difficulty in finding a package of the correct thickness and width. Felt weatherstripping may be applied by nailing, stapling, or gluing. (19)

RCR Limited manufactures quality weatherstripping called Climaloc. The Climaloc weatherstripping kits come in three color-coded series—yellow for good, red for better, and blue for best. In addition to the weatherstripping kits—which include an angle seal for the outer side of the door frame and either a doorsweep or threshold for the bottom of the door—RCR has a complete line of garage door weatherstripping and weatherstripping tapes.

Glass louvered windows can be made more airtight with the addition of special weatherstripping. Pemko's jalousie weatherstripping offers an extra flange that seals against the adjacent glass louver. The weatherstripping material can be tailored to fit easily, and it merely snaps into place. Unless you live in an area where there are a great many jalousie installations, your hardware dealer will probably have to order this type.

The best way to weatherstrip metal casement windows is to apply the special gaskets made for this purpose. They are easy to apply and usually quite effective. Frost-King Vinyl Casement Window Gasket is in wide distribution.

Interlocking metal weatherstripping systems require a more exact fitting than most other types. Therefore, the average homeowner should avoid installing such a system. A builder, however, may well opt for a version of the interlock when building a new home or adding on a new room. Pemko markets a wide line of interlock systems, including one for double-hung windows. Some window manufacturers even equip their products with such weatherstripping at the factory, but if your windows are not so equipped, the contractor or installer will have to cut a rabbet into the frame.(21)

Rigid strips designed to seal gaps around doors come in several styles, but they differ mainly in whether their spines are made of wood or metal and in the type of weatherstrip materials they hold. Mortell's Mortex Economy Doorstrip consists of closed cell vinyl on Ponderosa Pine. Mortell's Deluxe Doorstrip, Foamflex,

also has a spine of Ponderosa Pine, but its weatherstrip material is closed cell vinyl foam. Both are attached with nails. Mortell's Aluminum and Vinyl Door Weatherstrip, which is more expensive than either Mortex or Foamflex, can be used effectively on either wood or metal door frames. The wooden types, however, can be painted more easily to match the frame and be less obvious. (20)

Thresholds And Door Sweeps

The W.J. Dennis Fashion Line of door sealer products which includes door sweeps and door jamb insulation has recently added a replacement threshold made of vinyl instead of aluminum or wood. It is more attractive than most aluminum thresholds, and unlike its metal competitors the Dennis vinyl threshold forms a natural frost barrier. The plastic unit does not absorb moisture nor does it crack, chip or peel. Most importantly, the flexible vinyl insert seals against the door bottom for secure protection against drafts and energy-wasting air leaks.(22)

The Mortell's Automatic Door Bottom lowers snugly against the

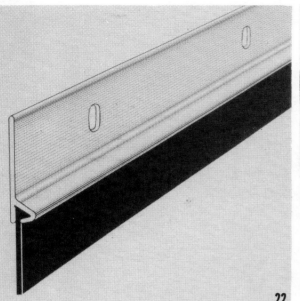

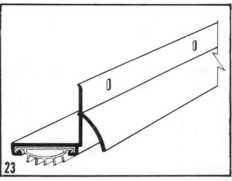

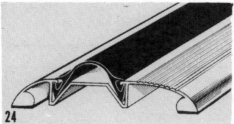

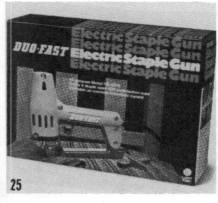

(19) Frost-King felt weatherstrip. (20) Mortex weatherstrip. (21) Pemko interlock weatherstrip system. (22) Fashion Line replacement threshold. (23) Pemko door shoe. (24) Dennis 200A replacement threshold. (25) Duo-Fast Electric Staple Gun.

floor when the door is closed, but it raises automatically to clear the carpet when the door is opened. A stop on the hinge side causes the Automatic Door Bottom to raise and lower at the proper times. The aluminum frame holds a flexible vinyl sealer strip, which does a good job of preventing air leaks under the door. Several manufacturers of thresholds and standard door sweeps offer variations on Mortell's Automatic Door Bottom.

Instead of a door sweep, the door shoe might be better suited to your needs. Pemko makes one of extruded aluminum with a fluted vinyl insert that provides a bind-free seal against either a metal or wooden threshold. Since the vinyl is under the door instead of in the threshold, it is protected from foot traffic. The unit's arched extension is a drip cap, designed to carry rain water that has run down the face of the door away from the bottom edge—thereby preventing water from coming in under the door. The shoe is mounted with screws through slots that allow for a degree of height adjustment. (23)

Metal door sweeps with vinyl inserts are becoming quite popular for sealing the gap between door and threshold. The metal frames can be attached easily with a few screws, and the slots in the frame permit a modest degree of height adjustment once the screw holes are made.

The most popular replacement thresholds are made of extruded aluminum and have replaceable vinyl inserts. The threshold is held to the floor with screws, and the vinyl strip is inserted last so that it hides the screw heads. The 200A threshold from W. J. Dennis and a similar unit from Macklanburg-Duncan are typical of the good aluminum thresholds available. There are also metal thresholds without vinyl inserts; these types usually require the addition of a door sweep to create a tight seal. You can also find wooden thresholds with vinyl inserts. (24)

Weatherstripping Installation Tools

Many types of weatherstripping require nails for installation. Although you can generally use any type of hammer, a magnetic tack hammer works best when driving small brads in cramped areas. Typical of the good magnetic tack hammers available is the Great Neck five-ounce model. Although well balanced and well made, it constitutes an in-expensive addition to your tool inventory.

If you have a choice between stapling or nailing weatherstripping, opt for stapling; it is much faster. The Duo-Fast Electric Staple Gun, moreover, makes stapling weatherstripping even quicker and easier. Electric staplers have long been used in industrial work, and now Duo-Fast has marketed a model for the homeowner. The gun accepts six of the most popular sizes in staple fasteners, and it can handle almost all of the stapling chores around the house in addition to plenty of small nailing and tacking jobs. (25)

While most of the packets of weatherstripping to be applied with nails come with an ample supply of the proper fasteners, you may find it more economical to buy your weatherstripping in bulk, and then you will need to buy your weatherstripping nails separately. Pemko markets such nails in all popular weatherstrip finishes—copper, bronze, cadmium, and stainless steel—although matching the proper nails to the weatherstripping is more important than shopping for any particular brand. Most retailers that sell weatherstripping in bulk also carry a line of weatherstripping nails.

Caulking

WHILE YOU CAN seal air leaks around the movable parts of doors and windows with weatherstripping, the passage of heated and cooled air from inside your house can still occur through cracks where two sides or surfaces meet and where two different types of building materials meet. The latter constitute "moving joints," i.e., the joint expands and contracts due to the fact that different materials expand and contract at different rates when subjected to changes in temperature, moisture, or pressure.

There are so many of these places in the average home that just a tiny crack at each one could add up to a rather large opening—possibly the equivalent of a hole in the wall one foot square. These trouble spots not only waste energy, but they also can lead to other damage in your home. They can, for example, allow moisture to enter, which may then cause rot and other problems. These cracks can also serve as entry ways for insects and other pests.

In most cases, caulking can solve the problem.

Caulk comes in several forms. The most popular—due to its ease of application—is certainly the cartridge, which is designed to be inserted in a caulking gun. Caulk also comes in squeezable tubes as well as in cans for application with a putty knife; the latter are called knife grade caulks. Another type is called rope caulk because it consists of strands of caulk packaged in a roll.

The most economical way to buy caulk is, of course, in bulk. The compound is then transferred from the bulk can into what is called a full barrel gun. Unfortunately, buying caulk in bulk is more economical only for industrial users; since the shelf life of most caulk is no more than about a year, the average homeowner should buy only what he or she will use in the near future.

Types Of Caulk

UP TO JUST a few years ago, there were no more than a handful of caulks and sealants on the market. Now there are many, and new ones will be forthcoming as homeowners realize the necessity for weatherproofing their houses.

Oil Base. The least expensive type of caulk currently available, the oil-base compounds adhere

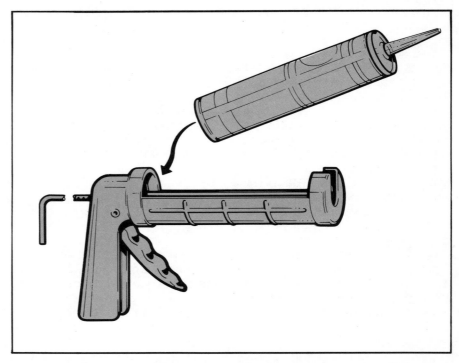

The most popular form of caulk is certainly the cartridge. Inserted easily into an inexpensive caulking gun, the cartridge makes it simple to apply caulk properly wherever it is needed.

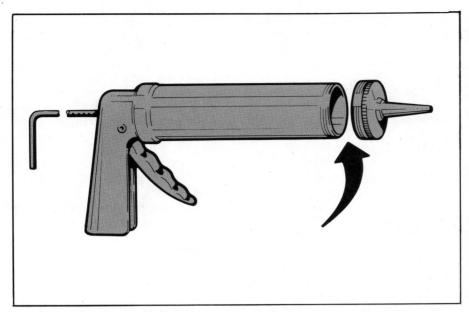

Caulk purchased in a bulk can must be transferred into what is called a full barrel gun. Although bulk cans constitute the most economical way to buy caulk, the savings are generally realized only by industrial users who use up the cans before the caulk's short shelf life expires.

to most surfaces, although they are best used on porous surfaces after priming. Applied to un-primed wood, the oil-base caulk can cause a stain. Most importantly, though, the oil-base caulks are not for use on moving joints. As the oil dries out, the caulk shrinks, thus losing its effectiveness in a relatively short amount of time. Under ideal conditions, oil-base caulk can last up to seven years, but under more typical circumstances it may not be effective beyond one year. You can prolong its life by painting the caulk after it dries tack-free. Use paint thinner to clean up any excess caulk.

Latex Base. Latex caulk also shrinks some, but less than oil base. Most grades can be counted on to last up to 10 years, and some manufacturers are now guaranteeing their latex caulks for 20 years. To compensate for shrinkage, apply a big bead of the latex caulk and use it to fill only narrow cracks. It adheres well to most surfaces, although metal surfaces should be primed or painted for optimum adhesion. Porous surfaces should also be primed. Since it dries in

15 to 30 minutes, latex caulk can be painted almost immediately. If used outside, latex caulk must be painted. It accepts almost any type of paint, but follow the manufacturer's suggestions as to required curing time if you plan to apply an oil-base paint. Latex caulk can be cleaned away with a damp cloth before it cures, but after that it must be cut or peeled away.

Butyl Rubber. Properly applied under the right conditions, butyl rubber caulk should last for ten years. It does shrink a little, but you can compensate for that by applying a wide bead and by using it only to cover small cracks. Among the lower cost caulks, it is not good for moving joints. On the other hand, it can be applied indoors or out, and it is often the choice for use on metal or masonry. Butyl rubber caulk can be painted with any type paint, but you should let it cure for a full week before painting. Many manufacturers suggest painting when their butyl rubber caulk is applied outdoors, but painting is not required. The caulk will dry tack-free in 30 to 90 minutes and can be cleaned up with paint thinner.

Polyvinyl Acetate (PVA). This sealant has fallen out of favor because it loses its flexibility when it dries. Use it only for filling small holes indoors. Even for such a limited purpose, moreover, polyvinyl acetate sealants are surpassed by others.

Silicone Seal. An excellent sealant which can last for at least 20 years, silicone seal adheres to just about any surface and shrinks very little. In addition, since it stretches up to seven times its cured width, silicone seal is ideal for moving joints. It dries tack-free in an hour and cures in two to five days, although a primer should be used before applying the sealant to porous surfaces. Some manufacturers specify that their silicone seal not be painted. Therefore, it's best to paint first and then apply such sealants; some come in colors. Sometimes you can paint a non-paintable silicone sealant, though, by first coating a cured bead with contact cement. The adhesive not only sticks to the silicone but also accepts the paint in many cases. To clean up after applying silicone seal, use paint thinner or naptha on porous surfaces or just a dry rag on other surfaces as long as the sealant is still wet. After it dries, you must cut any excess away.

Nitrile Rubber. Although a long-lasting caulk (life expectancy from 15 to 20 years), nitrile rubber shrinks considerably. Consequently, it should not be used on moving joints or wide (⅛ to ¼ of an inch) cracks. It adheres well to metal and masonry and other unprimed surfaces, but not so well to painted materials. It is extremely good for high moisture areas. It dries tack-free in 10 to 20 minutes and can be painted at that point, although painting is not necessary for protection of this compound.

Neoprene Rubber. This long-lasting compound has a 15 to 20 year life span and is especially good for use on concrete walls and foundations. Suffering only moderate shrinkage, it can be used on moving joints of no more than ¼ of an inch. Neoprene rubber becomes tack-free in about an hour, but it takes from one to two months to cure fully. It will accept paint, but paint is not needed for protection. Insofar as cleanup is concerned, ordinary solvents are of no help in removing neoprene rubber; use either MEK or tolulene, but be sure to read the caution notices beforehand.

Polysulfide. Capable of sealing moving joints for longer than 20 years, polysulfide compounds are thoroughly shrink resistant. On the negative side, though, these compounds are difficult for the novice to work with and they are toxic until cured—up to three full days before becoming tack-free. Furthermore, a special primer is required before a polysulfide compound can be applied to porous surfaces such as wood or masonry. Each manufacturer specifies the preferred primer, and a particular polysul-

fide usually does not adhere well to primers other than the one specified. Tolulene and MEK will clean away any excess.

Hypalon. Although not in general distribution, hypalon is an excellent caulk made to last up to 20 years. It can be used on moving joints and will adhere to any surface, but it requires priming before use on porous materials. While outperforming most other top quality caulks, it is no more expensive. Also in its favor is the fact that it's easy to work with.

Polyurethane. Polyurethane can be used on moving joints, will last up to 20 years despite weather conditions, and is easily applied. No priming is needed, but it does require 24 hours to become tack-free and up to two weeks to cure fully. Like hypalon, polyurethane is an excellent caulk that currently is not in as many retail outlets as it should be. Use paint thinner or acetone for cleanup.

Rope Caulk. Although it will last for a year or two, rope caulk should really be considered only a temporary solution to air leaks. It never forms a permanent attachment, but is just pressed against a surface and stays there until peeled away. On the other hand, it can be used on wide gaps (just peel off several strands), it does not dry out, and it does not shrink. Many people use rope caulk as a seasonal sealer around storm windows. It cannot be painted.

Where Should You Caulk?

AS A RULE of thumb, caulk any part of your home in which two different parts come together with a crack in between, and be especially vigilant to caulk places where two different building materials come together. Here is a list of spots to caulk.

1. Where the frames of doors and windows meet the sides of the house.

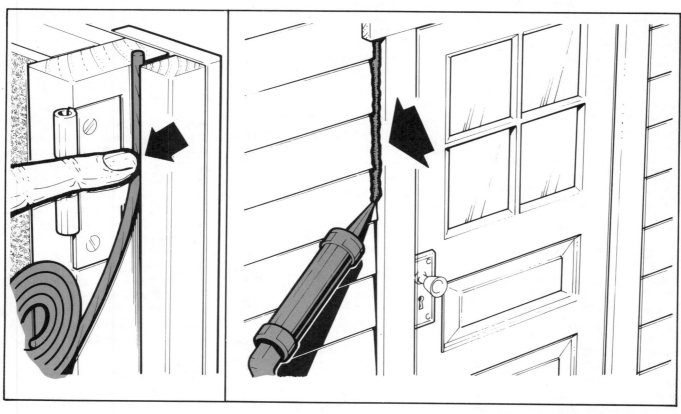

Rope caulk can serve as a handy temporary solution to air leaks. It never forms a permanent bond.

Be sure to caulk wherever the frames of doors and windows meet the sides of the house. As a rule of thumb, caulk any part of your home in which two different parts come together with a crack in between.

2. At the point where the sides of the house and the foundation meet.

3. In the joints where the steps and/or porches meet the main body of the house.

4. Where the chimney meets the roof, around the flashing, and in the gap in the seam between the flashing and the shingles.

5. Where plumbing pipes go through walls to enter the house.

6. Along the corner seams formed where siding meets.

7. Around the hole in the wall through which the exhaust vent for the clothes dryer goes.

8. In the spaces between window air conditioner units and window frames.

Caulking With A Cartridge

ALWAYS READ the manufacturer's instructions on the package of caulk. Different caulks have different characteristics that affect the way you should apply them to achieve top results. In general, though, you can follow these tips for gun-and-cartridge caulking.

Step 1. Always clean—i.e., scrape, peel, gouge, or pull—away all the old caulking. Once you get rid of all of it, clean the area to be caulked with a solvent. You want the area to be as free as possible of dirt, oil, and wax.

Step 2. Most caulks go on best when warm. Therefore, you should try to do your caulking work in warm weather. If that is not possible, warm the caulking tube itself before you apply its contents. On the other hand, most caulking gets too runny in extremely hot weather. In such instances, try placing the tube in the refrigerator for a brief period to slow down the caulk.

Step 3. Cut the spout at an angle and at a place that will give you the proper size bead for the job. The bead must overlap both surfaces to either side of the crack.

Step 4. Hold the gun at a 45-degree angle in the direction of your movement.

Be sure to caulk wherever plumbing pipes go through holes in the walls to enter the house. Be especially vigilant to caulk areas where two different materials come together.

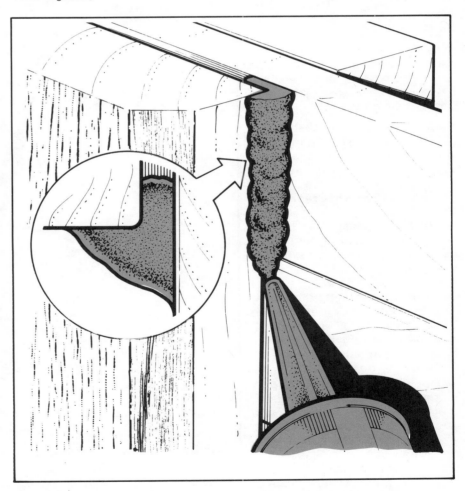

Make certain that the bead of caulk coming out of the tip of the cartridge is of the proper size for the job. The bead must overlap both surfaces to either side of the crack.

Cut the cartridge spout at the proper spot and angle for the job.

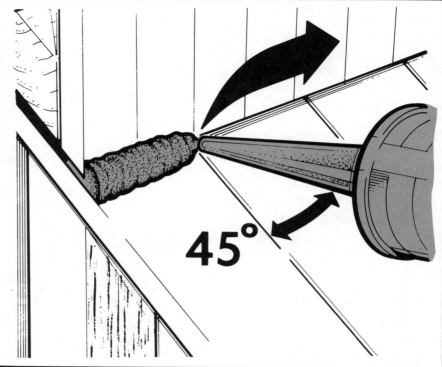

Hold the caulking gun at a 45-degree angle in the direction of your movement. After the bead is applied, you can generally smooth the caulk and push it into place with your finger to close a gap.

Step 5. Most caulks can be "tooled" (smoothed and/or pushed into place to close a gap) after the bead is applied. Some types can be tooled with a wet finger, while others can't be tooled at all. Read the label on the caulk for this information.

Step 6. In order to prevent the caulk from oozing out of the tube when you have to stop, twist the L-shaped plunger rod on the gun until it disengages.

Step 7. If you want to seal the tip of the tube, insert a fat machine screw in the nozzle and turn it in tight.

What About The Big Cracks?

GAPS THAT measure an inch or more in width cannot be bridged by caulking compounds. Moreover, some gaps are so deep that you could pump an entire tube of caulk in without filling the cavity. As a rule, any crack more than ½ of an inch wide or ½ of an inch deep requires something besides caulking compound.

Oakum, a treated hemp rope, available in plumbing supply stores, is often used to seal big cracks. Fiberglass insulation and sponge rubber strips can act as effective crack fillers, leaving a crack that can be closed with a double bead of caulking compound. In some cases, a piece of wood molding can be added in a corner of a wide crack, leaving only a tiny seam to caulk.

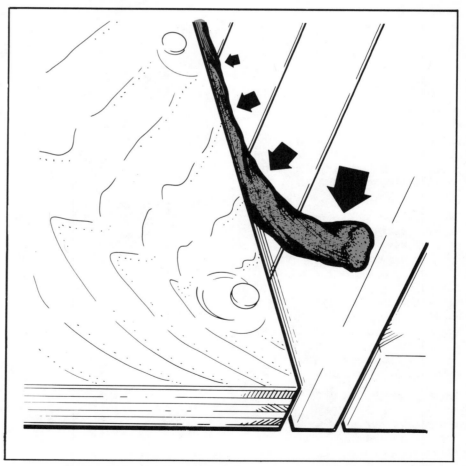

Oakum is a treated hemp rope (available in plumbing supply stores) that is often used to seal big cracks. Most cracks measuring more than 1/2 inch wide or 1/2 inch deep should not be filled with caulk.

Caulking

Materials and Supplies

Under the brand names of Seamseal and Polyseamseal, Darworth markets four different types of latex caulking compounds. Polyseamseal, an all-purpose latex caulk that comes in tubes as well as in cartridges represents the top of the line. It comes in white, gray, black and mahogany colors. Seamseal Acrylic Rubber Caulk comes next, followed by a vinyl acrylic based formula called Seamseal Latex Painters Caulk and finally by Seamseal Economy Latex Caulk. (1) (2)

Specialized Caulks

While not a true caulk, Polycel One, made by Coplanar Corp., is nonetheless a fine energy-saving product. A polymeric foam dispensed like shaving cream, it expands and forms a closed-cell rigid foam to fill cracks and spaces. Best used during construction to fill the gaps around door and window frames, Polycel One seals beneath floor plates and can also be foamed around plumbing pipe entries; in an older home, it can fill many gaps that would prove difficult to seal any other way. For interior use only, Polycel One is a new product and may be difficult to locate in many retail stores. Check with a local building supplies outlet.

Frost-King markets a popular brand of rope caulk in three sizes: Economy Pack, approximately 15 feet (enough for one average window); Bonus Pack, 30 feet; and Giant Pack, 90 feet. It is a quality brand, but like all rope caulks it does not form a permanent bond and has a short life span.

Vulkem brand urethane caulking compound, a product of Mameco Inc., is a good all-around caulk. It goes on easily; it can fill big cracks that would otherwise have to be stuffed with oakum; it adheres to any sound surface; and it accepts latex or oil-based paints. In addition, Vulkem's elasticity makes it good for sealing moving joints.

Complete Caulk Lines

Space Age caulks from GX International come in cartridges that need no gun. The self-dispensing tubes feed out the caulk with a twist of the wrist.

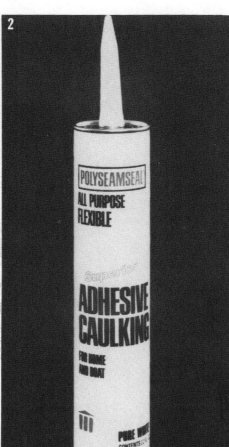
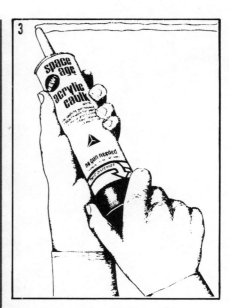

(1) Seamseal Economy Caulk. (2) Polyseamseal Adhesive Caulking. (3) Space Age Acrylic Caulk.

Most importantly, however, the Space Age caulks themselves—a complete line of many different types—are all high quality products. (3)

Contech makes quality caulks and packages them in tubes bearing large, bold designations. On the back of each tube, moreover, are easy to understand instructions and suggested applications. (4)

DAP makes just about all the different types of caulks that are commonly used by the homeowner. Easy to find in both paint and hardware stores, the caulks run from Rely-On oil base up to Butyl-Flex, which carries a 20-year guarantee. DAP also markets rope caulk and a couple of inexpensive caulking guns too.(7)

Another mass merchandiser distributing caulking products in hardware and paint stores throughout the country is the Red Devil Company. The Red Devil line includes latex, acrylic latex, butyl and oil-based caulk. All are popularly priced and of good quality. The same can be said of the Red Devil caulking gun. (5)

Painting silicone caulk has often proved a frustrating task for the do-it-yourselfer. But Dow Corning makes a paintable Silicone Caulk and Sealer that renders a tack-free seal that accepts either oil-based or latex paints. Best of all, this caulk carries a 20-year guarantee not to crack, crumble or dry out. (6)

Ruscoe's Permanent-Sealer is an all-purpose, nitrile rubber based sealant that can come in handy for caulking around aluminum frames and other metal objects. It comes in an aluminum color that cures to a natural aluminum finish as well as in white. A top quality product that can even be applied under water, Permanent-Sealer is available in both tubes and cartridges.

In addition to making a line of paint specialties, Nankee markets several types of caulks: e.g., Acrylic Latex Caulking Compound made from 100 percent pure acrylic resin; oil-base Elas-

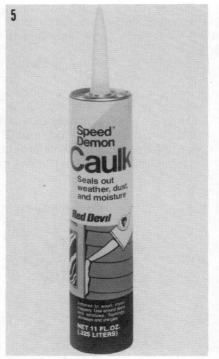

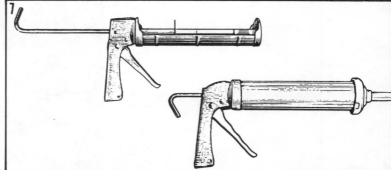

tic Caulking Compound that carries a five-year guarantee; and the budget-priced Star Caulk. All are comparable to the better known, more widely distributed brands.

Latex Caulk

Elmer's Acrylic Latex Caulk from Borden Chemical is a good product, and like all the other products in the Elmer's line, it is so widely distributed that you'll find it not only in hardware and paint stores but even in some large drug and variety stores. (8)

Properly applied, latex caulk should last at least ten years. Super Caulk from the Macco Adhesives Division of Glidden Coatings and Resins, however, is an acrylic latex caulk that carries a 20-year guarantee. Macco also markets a butyl caulk called Guard House that, like Super Caulk, is sold in both single cartridges and six packs. (9)

Caulking Guns

Having a good caulking gun around the house is a must. In addition to applying caulk, the caulking gun is a necessity for using the multitude of household compounds now packaged in cartridge form. For example, Michlin Chemical Corporation's ThinZit compounds in cartridges include concrete crack fix, roofing compound and a variety of caulks. Michlin also markets a good quality, popularly priced caulking gun.

While the average homeowner needs no more than an inexpensive caulking gun, Albion Engineering Company makes a costly version called the Cadillac of caulking guns. Model 139-3 features a 1/4-inch square piston rod with a thumb-activated instant pressure release. The gun's stroke can even be regulated to fit the size of the user's hand. Although an indisputably fine caulking gun, the Albion 139-3 should be considered for purchase only by homeowners who can use it on a very frequent basis.

(4) Contech caulks. (5) Red Devil Speed Demon Caulk. (6) Dow Corning Paintable Sealant. (7) DAP caulking guns. (8) Elmer's Acrylic Latex Caulk. (9) Guard House Butyl Sealant.

Storm Windows, Storm Doors, and Thermal Panes

THE DESIGNATION "storm windows" is really a misnomer because it implies that these windows are only for protection against the ravages of a storm. The term "insulating window" defines them better. Storm windows insulate an air conditioned home against outside heat during the summer, and they keep out the cold while keeping the heated air in during the winter.

No matter what you call them, storm windows or insulating windows are a good investment for most homes. Their energy-saving function, of course, is all the more obvious and crucial in climates of extreme heat and extreme cold where both air conditioning and heating systems are frequently utilized throughout the year.

In addition to their energy-saving function, insulating windows maintain a more uniform temperature within the home, thus making it more comfortable. When properly installed, moreover, they eliminate sweating on the primary windows, thereby reducing the frequency with which window sashes, window trim, and surrounding walls must be repainted. Insulating windows also are helpful in reducing the amount of outside noise, dust, pollen, and pollutants that enters the house.

But the primary reason people install insulating windows is that regular windows are responsible for 30 to 50 percent of a home's total energy loss. Some of this loss occurs around the frame; a problem which must be treated with proper caulking and weatherstripping. Much of the loss, however, occurs via conduction and radiation through the glass. When the air inside is warmer than the air outside, heat is conducted right out of the house through the window; in the summer, the situation is reversed. In the summer, moreover, the sun's rays are radiated through any glass it strikes directly or is reflected onto, causing the interior temperature to rise and the air conditioning system to work harder than it should.

By adding a second layer of glass, you can cut the heat transfer through conduction by about 50 percent. Another important factor in adding an insulating window is the creation of dead air space. Air itself can act as an insulator, reducing heat transfer through conduction.

You can have storm windows made to order and installed, buy them and install them yourself, or make them on your own from components available at large hardware stores, home centers, and lumberyards. Price out the aluminum sash, the corner locks and other hardware, the glazing channels, and the glass to see if doing it yourself is worth the effort. In addition, consider using plastic instead of glass; acrylic plastic sheets are available in various sizes at building supply houses.

If you decide to make your own storm windows, here's the way to go about it.

Step 1. Measure carefully the area where the storm windows will go.

Step 2. Cut (or have cut) the glass or plastic panes. The panes should be cut 1⁄16 inches smaller than the outside measurement of the frame, and they must be cut square to fit properly.

Step 3. Remove the rubberized glazing channels from the aluminum sash.

Step 4. Use a sharp razor blade to miter the corners of the channels, but don't cut through the outside edge. Merely cut out a triangle so that the miter can be made where the channel edges will be whole except where the two ends meet.

Step 5. Slip the channels around the edge of the pane, use tape to hold them in place if necessary.

Step 6. Use a hacksaw and miter box to miter the aluminum sash pieces.

Step 7. Install the corner lock hardware in the two side pieces.

Step 8. Fit the top and bottom sash pieces over the glass and glazing channel. Then install the side pieces.

Step 9. Attach adhesive-backed foam weatherstripping all the way around the inside edge of the frame to seal out air leaks.

44

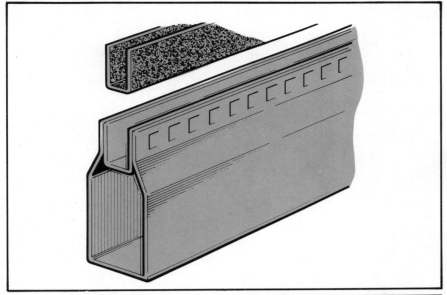

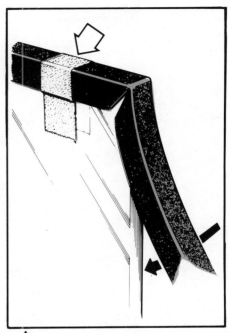

◄ Before you cut the aluminum sash pieces when making your own storm windows, be sure to remove the rubberized glazing channels.

▲ After you miter the corners of the channels with a sharp razor blade, slip the channels around the edges of the pane and secure them in place with tape if necessary.

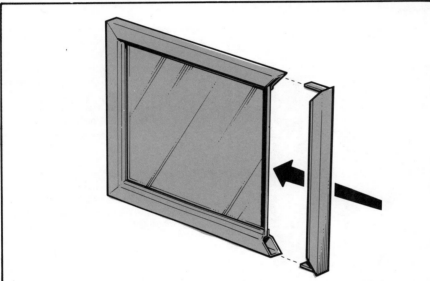

◄ Fit the top and bottom sash pieces over the glass (or acrylic sheet) and over the rubberized glazing channel. Then install the side pieces.

Installing your own storm windows, whether you make them or buy stock sizes, is not a difficult task. In fact, in most cases it is a very simple one. The storm windows must be flush with the outside frame of the primary window, and the air space between should be from ¾ of an inch to 1½ inches. Most frames have a bottom expander that allows you to adjust the height so that the storm window fits properly.

You can mount the storm windows with special brackets, or you can—if you intend to leave the storm windows up year-round—attach the additional windows to the frame of the primary windows. Just drill holes through the storm window frames, and attach them to the primary window frames with screws.

If you opt for the permanent attachment method, run a light bead of caulk along the top and sides of each storm/primary window combination. And be sure to test the sashes of the primary window to make sure they can move before tightening the storm window screws. If the sashes don't move freely, the storm window frame is probably out of square and must be corrected before tightening the screws fully.

If you decide to have someone else make and install your storm windows, check out the seller/installer with the Better Business Bureau, with other consumer agencies, and with some of the firm's recent customers. Then make certain that the insulating windows actually installed on your house are of the same quality as you were shown in the sales presentation.

The key factors that determine storm window quality are the gauge of metal in the frame, the finish, the corner joints, and the seal around the pane of glass. The heavier the gauge of metal, the sturdier the window unit. Some lightweight frames bend so easily that they fail to last more than a year or two.

Unfinished aluminum will oxidize and pit, presenting you with an unwelcome maintenance

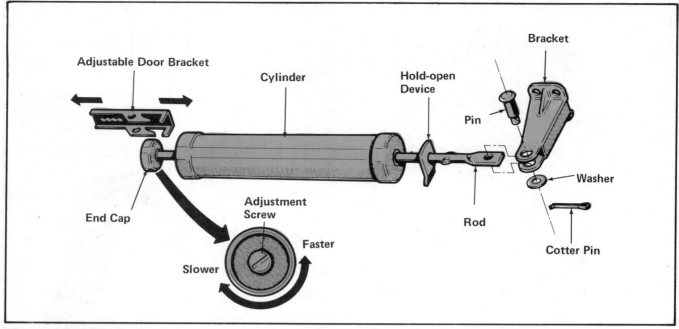

Adjustable Door Bracket

Cylinder

Hold-open Device

Bracket

Pin

Washer

End Cap

Adjustment Screw

Faster

Slower

Rod

Cotter Pin

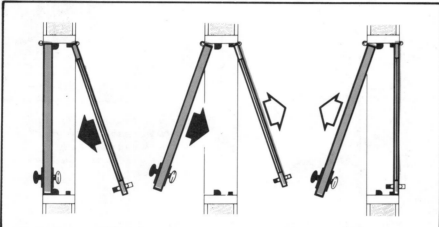

▲
Most storm doors come equipped with automatic closers that prevent an entry way from ever being totally open for very long. Good closers have adjustment screws that allow homeowners to regulate the rate at which the door closes as well as devices to hold the door open.

◄ *Since storm doors open out while entry doors open in, the storm door helps prevent inside air from escaping or outside air from entering. Seldom are both doors open at the same time, and if the storm door is equipped with an automatic closer, that brief interval is minimized.*

chore. Anodized or baked-on enamel finishes never require any attention.

The corner joints must be strong or else the window unit will not maintain its shape and thus won't fit properly. Overlapped joints are usually preferable to mitered joints.

Usually, you can tell by examining the frame carefully whether the glass and the weatherstripping are set solidly in an adequate groove. If the windows have movable inserts, be sure to check them for ease of movement.

It's always wise to get several estimates before contracting to have storm windows made and installed for you. Both prices and delivery time can vary greatly.

Storm Doors

WHILE DOORS do not conduct heat as readily as the glass in windows, they do have a conduction factor that can cost you energy dollars. Outside doors can also permit air infiltration, another way that your home's energy can be wasted.

On the other hand, the cost of a storm door is considerably higher than that of a storm window, and since the energy loss is

much less through a door, it will take much longer to recover your investment. Nonetheless, there are many cogent reasons for adding a storm door. Most have an automatic closer which means that the entry is never totally open for long. In addition, since storm doors open out while entry doors open in, the storm door helps save energy in the opening process. This means much less air is exchanged between inside and out.

The storm door also acts as a deterrent to burglars. Steel doors with ornamental iron grills and sturdy deadbolt locks give your home great extra security.

Like storm windows, moreover, storm doors reduce the amount of outside noise, dust, pollen, and pollutants that would otherwise enter your home. They also protect primary doors and thus minimize the amount of maintenance these doors require.

Buying an inferior storm door in order to save money is a terrible investment. Unlike windows, doors are moved frequently and are subject to many more problems. A door made of thin gauge metal can warp, sag, and bind, and when the door no longer fits tightly, infiltration will again be a problem. Make sure that the door you buy is equipped with either tempered glass or an acrylic plastic of adequate thickness for safety. The extra money spent on a good door will pay off in longevity, whereas a cheap door may have to be replaced after only a year or two.

In most cases, the average homeowner would be wise to have a new storm door installed by a professional. A skilled do-it-yourselfer, though, can perform the installation with ordinary tools and a friend to help hold the door while it is being installed.

Thermal Pane Windows

UNLESS YOU are building a new home or putting on an addition to an older home, you may not be able to utilize the advantages offered by thermal pane or multi-glazed windows. And those advantages—from the energy standpoint—are considerable. Double-glazed windows match the efficiency of storm sashes in reducing energy loss by about 50 percent, and triple-glazed thermal panes hike that figure to about 65 percent.

In thermal pane windows, the space between the two or three panes contains dry air or gas hermetically sealed to eliminate condensation. The dry gases are better than the dry air in that they provide higher insulating efficiency. Thus triple-glazed windows with dry gases between the panes offer truly excellent energy efficiency.

Multi-glazed windows come in all styles and sizes—sliding, casement, double-hung, etc., with either wood or metal frames. Wood frames do not conduct heat as readily as metal ones do,

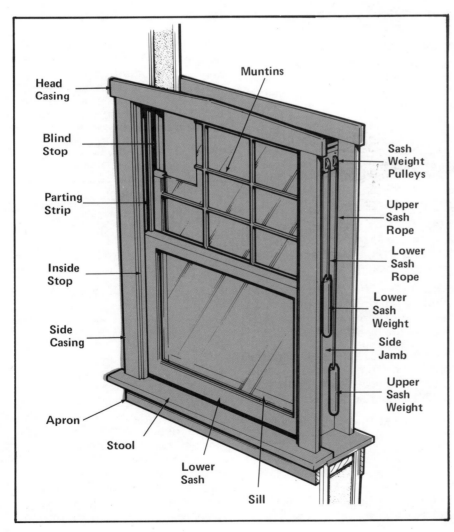

▲
You can mount storm windows with special brackets or attach them to the frames of the primary windows. Just make sure to leave an air space of from 3/4 of an inch to 1-1/2 inches between the storm and primary windows; the air space serves as insulation.

◄ *A typical double-hung wooden window can usually be replaced with no structural alteration and frequently in a fairly short amount of time. In fact, window replacement can be regarded as a do-it-yourself task by the homeowner who has some experience in home repairs.*

and wood has about three times the R-value of aluminum—the most common metal used for window frames. If you opt for wood, though, make certain that the frames have been properly treated with a good preservative.

The idea of replacing windows in an existing house may sound impractical, but in most cases windows can be replaced with no structural alteration and in a fairly short amount of time. In fact, window replacement constitutes an endeavor that many homeowners tackle as a do-it-yourself project. It is not, however, a project for someone who lacks experience in performing basic home repairs.

In brief, here are the steps involved in replacing a standard double-hung wood window.

Step 1. Measure very carefully the opening that the new window is to fill. You may find that a stock size will fit, but it is better to have the new window made to order if the stock size isn't quite right. Since the prices of stock and made-to-order windows are similar, the only disadvantage is the delay in getting a custom-ordered window. Shop around for prices and for delivery estimates. The window unit—whether made-to-order or stock

When prying out the inside moldings or stops, be sure to work carefully. You'll have to replace these.

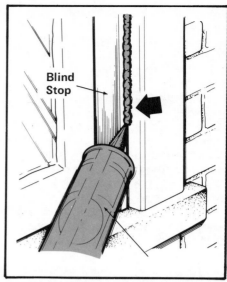

After you've installed the replacement window, be sure to caulk every crack where outside air could enter.

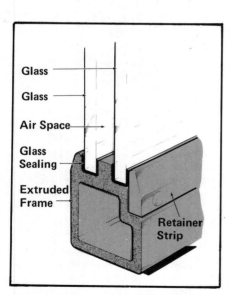

The multi-glazed replacement window comes as a complete, preassembled unit whether it is custom ordered to fit an unusual size or a stock window available at large home centers.

—comes as a complete, pre-assembled unit.

Step 2. Pry out the inside moldings or stops very carefully, you will replace them later.

Step 3. Remove the ropes or chains and the weights and pulleys from the double-hung sash. If the weights are inaccessible, let them drop into the frame cavity.

Step 4. Take out the old bottom sash.

Step 5. Remove the parting strips so that you can take out the upper sash.

Step 6. If the new window has removable sashes, take them out to make the unit lighter and easier to handle.

Step 7. Test the fit of the new window unit in the opening. You can compensate for minor misalignments by adding shims.

Step 8. Remove the new window from the opening, and run a bead of caulking around the blind stop at the top and both sides.

Step 9. Seat the unit against the caulked stops, and then secure it with screws.

Step 10. Most window units come with alignment screws and an expander header to make the unit fit square and snug. Adjust

these components as indicated.

Step 11. Caulk around the inside frame of the unit before replacing the inside stops.

That's all there is to it. Of course, you may encounter other steps, depending on the type of replacement and the condition of the wood around the opening, but the preceeding represents a fairly typical installation.

Tinted Glass Windows

TINTED GLASS absorbs more heat than does clear glass, leading many people to think of it as a great energy saver. Since the absorbed heat is reradiated on both sides of the glass, however, not all of the absorbed heat is diverted in the desired direction. That is not to deny the advantages tinted glass offers, advantages that may or may not be important to individual homeowners.

Tinted glass is often used commercially to reduce the amount of light entering a room, thereby contributing to the enhanced visual comfort of the occupants. It can achieve such a reduction while still maintaining an adequate level of light in most cases. Many times tinted glass is used strictly to prevent fading by controlling the amount of ultraviolet rays coming through windows; it does an excellent job along these lines.

If you are concerned about heat penetration but ultraviolet rays are no problem, you'd be better off installing reflective glass rather than tinted glass. The reflective glass will bounce the sun's rays back outside, although most types of reflective glass do not screen out ultraviolet rays.

Avoid the tint films designed for application to existing windows. Instead, apply a solar reflective film which does a much better job of energy saving. The solar film not only makes the

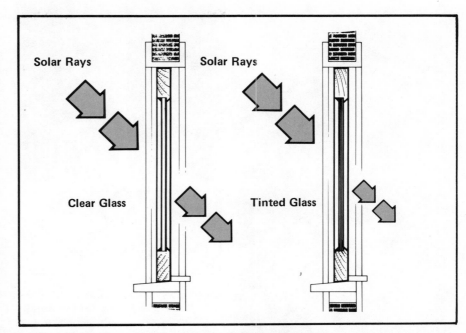

Tinted glass absorbs more heat than does clear glass, and it does an excellent job in reducing the amount of ultraviolet light entering a room. Reflective film, however, is preferable for heat reduction where ultraviolet light is no problem.

glass absorb more heat, but it also reflects heat back outside; it even does a better job of controlling ultraviolet rays. Since so-

lar reflective film and tinted film cost about the same, moreover, there is little reason to select the tinted film.

Storm Windows, Storm Doors, and Thermal Panes
Materials and Supplies

Storm Windows

Most storm window retailers carry a sample of each style, but they have to order the actual windows from suppliers who are frequently located in other cities or even in other sections of the country. The order, therefore, may not be filled for several weeks. You may find, however, that the large lumberyards and home centers in your locale stock several standard sizes of storm windows. Unless you have

odd-sized windows that require custom-made storms, you should be able to find what you need in stock and be able to get them up right away.

While they aren't as good as actual storm windows, you can create quick plastic windows by applying a special film on window screens. R-V Lite's Easy-Does-It kits include enough R-V Lite plastic film, fiber strips and nails to make one, two, or four window conversions. The transparent plastic film, which lets

light in as well as allowing you to see out, succeeds in forming a barrier between the cold outside and the windowpane. Though they cannot provide the insulating qualities of true storm windows, these R-V Lite conversions do help in eliminating drafts.

Plaskolite Inc. makes an easy do-it-yourself storm window kit that consists of sheets of clear rigid plastic and a softer trim material to fit around the edges of the sheets. The plastic sheets

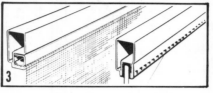

(1) In-Sider Storm Windows. (2) Season-all storm windows. (3) Reynolds aluminum components. (4) Reynolds interior storm window. (5) Climate Guard windows. (6) Grande Entrance door. (7) Storm King door closer. (8) Kinkead thermal door.

come in standard window sizes, but they can be cut easily. In-Sider Storm Windows are installed from inside the house on the inside of the regular window. Just put the plastic in place and apply the trim around the outside. The trim will hold the plastic in place as well as make it airtight. In-Sider Windows stay securely in place, but they can be removed for cleaning or when fresh air from the open window is desired. (1)

You can make your own storm windows by placing aluminum storm sash sections and aluminum corner locks around sheets of either glass or acrylic. Reynolds Metals Company makes all the aluminum components you need. Just measure the storm sash sections and saw the aluminum to size with either a fine-toothed hacksaw or a coping saw. Then fit plastic glazing channels around the pane, and install the sash sections and corner locks over the channels to create an airtight seal. (3)

Season-all storm windows are available in a complete line of double-hung, sliding, picture and basement window styles to

weatherproof practically any window. Available in mill or anodized finishes, or in white, black, or brown baked enamel, Season-all guarantees its windows against chipping, cracking, blistering and peeling for 15 years. (2)

Interior insulating panels for windows can provide about as much energy savings as standard storm windows. Reynolds Metal Company's line of do-it-yourself aluminum products now includes an interior storm window. It consists of an aluminum frame that supports a sheet of clear polyethylene film. Self-adhering foam tape provides an airtight seal and permits easy installation and removal. (4)

Storm Doors

The Metal Vent Manufacturing Company Inc. markets an excellent line of storm doors under the Better Bilt and Tailored Tuff brand names. The doors are made of heavy steel frames and are decorated with cast ornamental iron work; all have heavy-duty hardware and heavy tempered glass panels that can

be interchanged with screens. An optional heavy-duty, brass, double deadbolt lock is available. The 58 different designs are available in black or white, or they can be custom painted to specification. They offer the energy-saving qualities of storm doors and are attractive and secure.

A sliding glass patio door that is not closed all the way is an energy waster. The Gemini 10-M closer from Wartain Lock Company can eliminate the open door problem and thereby reduce energy loss. Easily installed, the 10-M automatically closes sliding doors. It can be adjusted to the weight of the door and has an additional adjustment to regulate closing speed.

Alsco Anaconda makes a wide range of aluminum storm doors, all of which feature an automatic closer, a sill expander to provide a perfect fit, safety glazing, and marine-type glazing cushions that allow for easy replacement of a broken pane. The doors come in a choice of three finishes—mill (the natural aluminum appearance), white and bronze—and in six styles. Some of the

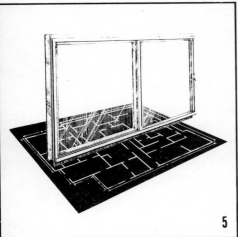

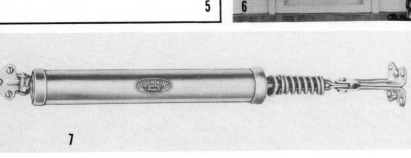

doors are equipped with self-storing panels to permit quick conversion from a storm door to a screen door.

Season-all's Grande Entrance is a new full-view storm door that features an uninterrupted 74-inch span of glass. The idea behind having such a storm door is to display nearly all of the main door behind it. Available in models with tempered safety glass that can be replaced by a full screen for warm weather ventilation, the Grande Entrance can be ordered in anodized aluminum finish or white, black or brown electrostatically applied baked enamel finishes warranted not to chip, crack, blister or peel for 15 years. Extruded corner gussets, spring-loaded Oilite bearing hinges and shock-absorbing marine-type U-channel vinyl glazing are some of the features of this premium quality storm door. (6)

An open door policy certainly is an energy waster. Storm doors are in particular need of a good, efficient closing device. S-B Manufacturing Company Ltd. markets the Storm King storm door closer that features adjustable closing speed, a cush-ion-spring shock absorber and a positive-locking device to hold the door open when needed. (7)

The Kinkead Division of U.S. Gypsum markets a complete line of thermal entrances, single and double doors plus matching sidelight panels designed for people who are building or remodeling. The doors feature a core of uniform density (no skips or hollow spots) polystyrene foam within a wooden frame. A skin of galvanized steel is flanged to the wooden stiles and rails. There are many different door styles in standard entry sizes. All are weatherstripped and equipped with an adjustable threshold. (8)

Thermal Panes

Pittsburgh Plate Glass markets thermal panes that can save homeowners a great deal of energy. Twindow Xi units consist of twin panes fastened together by a special welding process that seals the two pieces of glass completely all the way around. Sealed between the panes is a special dry gas that provides better insulating qualities than did the dead air space in old-fashioned multi-glazed windows. These special panes can be fitted into frames of either wood or aluminum in all window styles.

Nu-Prime Model 105 windows, made by Season-all Industries, Inc., are double-hung aluminum replacement windows that are double weatherstripped. Available in either single or optional insulating glass, Nu-Prime windows are custom made to the exact measurements of the opening, and yet they are competitively priced with stock windows. They can be installed in any type of opening—wood, brick or masonry.

Reynolds Climate Guard windows and sliding doors feature a thermal barrier that acts as a buffer zone between the outside metal frame and the inside metal frame, limiting the transfer of heat from one side of the metal frame to the other. This molded-in-place heat barrier, when combined with double glazing, substantially reduces heat transfer. Climate Guard windows and sliding doors can be used as replacement units or in new construction. (5)

Reflective Film

YOU HAVE probably seen windows in office buildings that look more like mirrors than clear glass. This reflective quality—achieved through the application of a bronze- or gold- or silver-colored film—has proven itself to be a big energy saver. It not only reflects the sun's heat away from the building during the summer, but it also reflects the heat created inside back into the room during the winter.

Now this same reflective film treatment is showing up on residential windows. In addition to providing summer and winter energy savings, the film greatly reduces sun-caused fading of drapes, carpets, and furniture. It cuts out glare to make the home more pleasant, and it converts regular window glass into a sort of safety glass that resists shattering. Finally, although it still lets light in, the film enhances privacy by making the glass almost one-way.

Some manufacturers provide sun control film in several degrees of reflectiveness. People who don't like the mirrored look can select a film referred to as 50 percent. It looks just slightly smokey, but it possesses enough reflective qualities to make a real difference. The intermediate version is called 65 percent, while the bronze films from some companies reflect up to 80 and 85 percent. The larger the percentage, the greater the energy-saving benefit.

Installing sun control film used to be considered a job that could be done only by a pro. Now, however, there are several brands of film that the homeowner can install with professional results. Installation procedures vary by brand, but the following step-by-step guide—based on the instructions for installing the popular Solar-X film—shows how simple the job can be.

Solar-X markets an inexpensive tool kit with everything you'll need for the installation. If you don't have the kit, you'll need a rubber squeegee, a spray bottle, a metal straightedge, a utility knife, and paper towels. If there are paint spots or flecks on your windows, you'll also need a scraper.

Here's how to apply the film. All work is done inside.

Step 1. Mix one teaspoon of liquid dishwashing detergent and one pint of water in a spray bottle.

Step 2. Spray the window with the solution and wipe it dry with a paper towel to be sure the window is clean.

Step 3. Cut the piece of reflective film so that it is roughly ¼ of an inch larger than the window.

Step 4. Put tabs of masking tape

Place short strips of masking tape on both sides of the reflective film at one corner, and don't permit the sticky sides of the tape to touch one another. The tape will help you separate the film from its backing.

on both sides of the film at one corner. Don't let the sticky pieces of tape touch.

Step 5. Spray the glass with water until it is completely covered.

Step 6. Pull the tabs of masking tape to separate the film from its backing.

Step 7. Spray the sticky side of the film with water as you peel away the backing. The entire surface of the film must be wet.

Step 8. Place the sticky side of the film against the glass, and slide the film around until it is positioned properly. You will find that the wet film slides easily against the wet glass.

Step 9. When the film is in position, spray the back side of it with water to provide lubrication for the squeegee.

Step 10. Work the water from the center of the film by making both vertical and horizontal strokes with the squeegee. Don't worry about tiny bubbles or haze; they'll disappear in a few days. And don't press the squeegee down along the film's edges until after you trim the film. Some people wrap a paper towel around the squeegee to absorb any surface water.

Step 11. Use the utility knife and straightedge to trim the film to within about 1/16 of an inch from the edge of the pane.

Step 12. Squeegee the edges of the trimmed film.

That's all there is to it. Now, just let the film cure for a few days, after which you can clean it with clear water or with a mild detergent solution. Do not, however, use ammoniated window cleaners, and never use anything but a rag, chamois, or squeegee for cleaning.

If the film ever starts to peel up along an edge, apply a thin line of clear fingernail polish to the problem area and press the film back down again.

Other brands of reflective film may require other installation methods. Therefore, be sure to follow the manufacturer's directions for the brand of film you plan to install.

When the inside of the window and the sticky side of the film are completely covered with water, place the sticky side of the film against the glass. Then slide the film around until it is positioned properly.

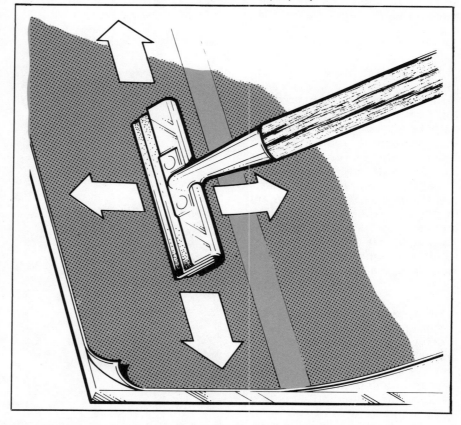

Move the squeegee in both vertical and horizontal directions to push the water out from the center of the film. Do not, however, press the squeegee down along the film's edges prior to trimming the material.

Reflective Film

Materials and Supplies

Homeowners interested in using a sun control film on windows to help reduce their heating and cooling costs might consider the film manufactured by Solar-X Corporation. It is available in silver or bronze and in several densities ranging from 50 to 85 percent. The 50 percent film does not produce the mirrored effect of the higher density film, making it more desirable to some people. Solar-X dealers offer a five-year written warranty on material. The film is also available at retail chains such as Montgomery Ward and J.C. Penney. Although the installation of the film requires but a few simple tools and many households have these tools on hand, dealers offer a kit containing all the tools necessary to install this film.

If you want to protect your furniture upholstery or drapery fabric from strong sunlight, reduce the harsh glare of the sun and help keep your home cooler in the summertime, you can apply a reflective vinyl covering to the inside of your windows. Tint-A-Pane, manufactured by Hartwig-Hartoglass, Inc., will remain in place and it can be cleaned just as if it were a pane of glass. To install, cut the vinyl material to fit the window, wet it, and smooth it on the pane with a sponge or squeegee.

Homeowners, building engineers and architects have been looking for ways to pocket energy savings. One method is to install sun control films to reduce air conditioning costs, to lessen the load on a building's cooling system and to prevent sun damage to drapes and furniture. Scotchtint Sun Control Film, made by 3M Company, is de-

signed to resist up to 75 percent of the sun's heat, 82 percent of its glare and 81 percent of its ultraviolet rays. All authorized dealers we checked, including Sears, who sell and install this product, offered a two-year guarantee against peeling, cracking, crazing or loosening when the installation is arranged by the dealer. In addition, some dealers do not sell the film without installation. This tends to discourage the do-it-yourselfer. If you do purchase sun control film

to apply yourself, the type with an "A" before the identification number indicates it is rewettable film; "P" indicates pressure-sensitive film. (1)

There are many sun control films on the market, but information has been limited only to top-quality products in nationwide distribution. First, you should know that there are two basic types. One is pressure-sensitive; it has a backing film

(1) 3M Scotchtint Sun Control Film.

that is peeled off. The other type is rewettable. Both types of film are very effective. The difference is the adhesive. If you have windows that "sweat" a great deal, for example, obtain the pressure-sensitive type of film. The rewettable film can loosen if you have a problem with moisture. In addition, if you live in an apartment, consult the landlord before installing either type of film. You may be able to convince him to install such film for the entire buildling; if he is going to want the film removed when you vacate, you will want to install the rewettable type, which is removed more easily. Rewettable film, unfortunately, is not usually available to the do-it-yourselfer.

The human body radiates heat, and infrared equipment can detect heat when a surface is scanned. This infrared photograph taken from outside a window shows a man holding a sheet of reflective sun control film in front of him. This graphic demonstration of the ability of solar control film to stop the transfer of heat was made available by Modern Aire Sun Control Products, Inc. (2)

To help keep your home warmer in winter and cooler in summer, to reduce fading of rugs, draperies and furniture and to maintain more privacy, you might consider Reflecto-Shield, a transparent window insulation film made by Madico. The product, available in gold, silver and bronze colors, reflects up to 77 percent of the sun's rays. (3)(4)

To save some fuel, you might consider the use of a window-tinting film on the windows of your auto, camper or trailer where it is permitted by law. One such poduct is Win-Do-Shield made by Madico, which also manufactures Reflecto-Shield Sun Control Film for home use. In warm weather, it can help prevent air conditioning equipment from putting an extra load on your engine and wasting fuel; in cold weather, it can help keep heat inside the vehicle. The product, however, cannot be applied to double curved surfaces.

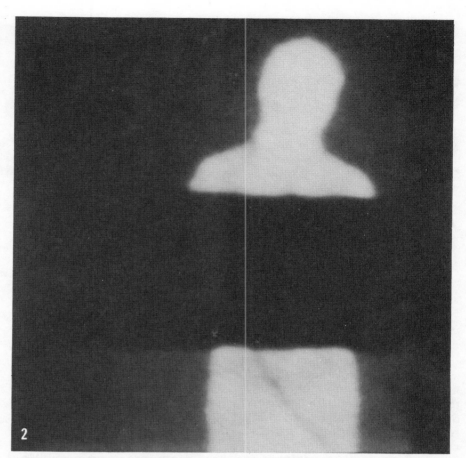

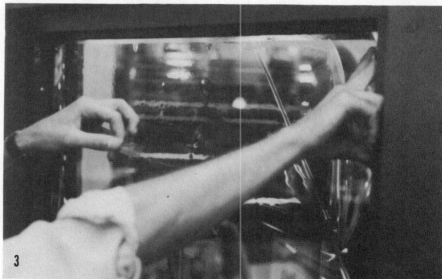

(2) Infrared photo from Modern Aire Sun Control Products. (3) & (4) Reflecto-Shield window insulation film.

Shades, Blinds, Shutters, Drapes and Awnings

REFLECTIVE film can be an effective energy saver during the summer because it prevents much of the sun's heat from entering the room, thereby easing the air conditioner's work load. It stands to reason, therefore, that materials capable of blocking out the sun altogether would constitute a good way to control heat. Shades, blinds, shutters, drapes, and awnings can all block out the sun.

Shades. Shades can reduce heat gain during the summer by nearly 50 percent. In addition, they can cut heat loss during the winter by up to 25 percent. As a result, shades can produce significant energy—and financial—savings. Blackout shades, moreover, are ideal for sealing out the daylight from rooms of late risers or mid-day nappers. Although shades used to be quite unattractive—and lost a good deal of their popularity for that reason—they are now available in a variety of colors, fabrics, and trim, making them decorative as well as functional window treatments.

Blinds And Shutters. Both of these window installations offer the same energy-saving qualities as shades do, but unlike shades, blinds and shutters can be adjusted so that they block out the sun while still permitting some light to enter the room.

Drapes. Fabric itself possesses insulating qualities that can keep cooled air inside during summer and heated air inside during winter. Even better, though, are the new thermal-backed drapery fabrics that prevent summer heat from coming in and winter heat from going out. Insulated drapery liners are also available. Energy-conscious homeowners should definitely look for insulated drapery fabric and liners when shopping for new or replacement drapes.

Measure windows from jamb to jamb when ordering shades that will be installed in inside brackets.

Awnings. Properly positioned window awnings can reduce heat gain from the sun by as much as 75 percent. Most effective when positioned over windows facing east or west (where they can combat direct sunlight), awnings are also very helpful in keeping the sun off a window air conditioning unit. In addition, awnings represent an inexpensive way to put a cover over a doorway, not only for sun protection but also as a shield against precipitation.

Available in metal, wood, and canvas versions, awnings can be stationary or movable (fold-up or roll-up). The most effective awnings are those that allow for air circulation as well as blocking out the sun. Since awnings can be detrimental during the winter—as they continue to block out the sun when it could be helping warm the house—energy-conscious homeowners should opt for awnings that can be removed or moved out of the way when the cold weather starts.

How to Measure, Install, And Adjust Shades

SHADES CAN BE installed in several ways. The brackets that hold the shades can go inside the window frame, outside the frame, on the frame itself, or on the ceiling. Different brackets are available to suit each of those different installation situations. If you want the shade to be within the window frame and the frame is deep enough to accommodate the roll, use regular inside brackets. When you want the frame to show but it's too shallow to let the shade clear the window hardware (or if the shade would look better being farther out), install inside extension brackets. When you want shades wider than the window, mount outside brackets on the wall or on the frame itself. Ceiling brackets are only slightly different than outside brackets.

There are special shades and brackets available for horizontal and for bottom-up shade movement. These unusual types work on a pulley system that can keep the shade taut at any point. If you face a situation in which either of these applications might be of help, ask your shade dealer about them.

After you decide where you want the shade, measuring is simple. Use a wooden or metal tape measure rather than a cloth tape measure, however, because a cloth tape can stretch. And, while two or more windows may look identical in size, it's best to measure each one to be sure you get shades of the right dimensions for every window.

For shades installed in inside brackets, measure from jamb to jamb and specify when ordering your shades that you plan to install them in inside brackets. For outside or ceiling or frame-mounted brackets, merely make light pencil marks where the brackets go and measure the distance from one mark to the other. With outside and ceiling brackets, the shades should overlap the inside edges of the frame by at least an inch on each side, and you should be sure to mention the type of brackets you are using when ordering the shades.

When replacing an old shade, measure the distance between the tips of the pins and specify when ordering the new shade that this is a tip-to-tip measurement. While most shade shops cut the width to order, shades do come in standard lengths. Therefore, before you place your order, measure the full height of the opening and then add from eight to twelve inches to take care of roll over.

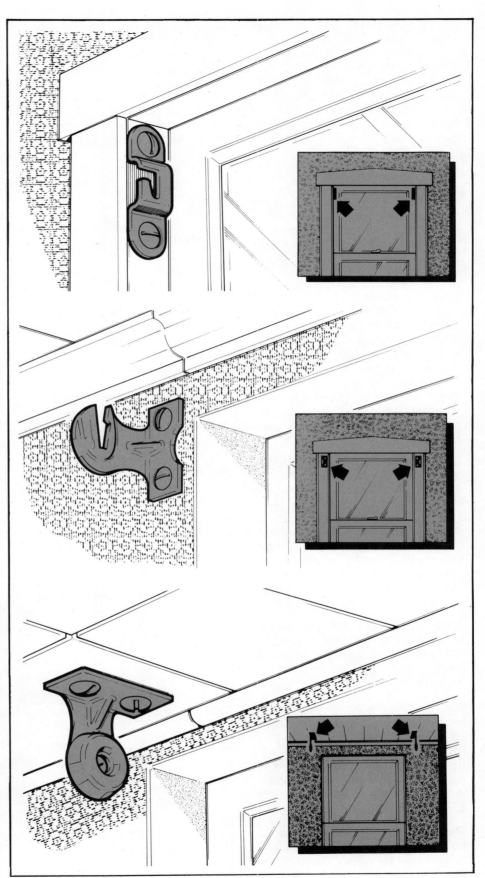

Use inside brackets (top) when you want the shade to be within the window frame, outside brackets (middle) when you want the shade to be wider than the window, or ceiling brackets (bottom) when you want to cover the wall area above the window frame.

Installing shades is quite simple, but you must make certain that the brackets are secure. Brackets installed in wood can be put in with nothing more than wood screws. If the brackets are to be installed in sheetrock or plaster however, you should use plastic wall anchors along with the screws.

How To Adjust Shades

IF A SHADE refuses to operate properly, you can make some simple adjustments. The following steps should help you cure most of the common shade woes.

Step 1. A shade that goes up too fast and with too much force is telling you that there is too much tension in its spring. To relieve this excess, roll the shade all the way up, remove the flat pin end, and unroll the shade about two revolutions. Replace it and see whether you have solved the problem. If not, repeat the process again.

Step 2. A shade that won't go up lacks sufficient spring tension. Pull the shade down and remove the flat pin end from its bracket. Now roll the shade back up about two revolutions by hand and put the pin back in its bracket. Test the shade. If it still refuses to go up, repeat the spring-tightening process until there is enough tension to move the shade.

Step 3. If the shade wobbles as it moves up or down, a pin is probably bent. Remove the shade and gently straighten the pin with a pair of pliers.

Step 4. If the shade falls out of its brackets, the brackets are either loose or too far apart. Tighten or reposition the brackets as indicated. If the brackets are mounted inside the window casing, you must add cardboard shims behind them.

Step 5. A shade that binds is being caught by brackets that are too close together. Reposition the brackets to allow for proper shade movement. If the brackets are mounted inside the window casing, try tapping the brackets lightly with a hammer.

Step 6. If the shade won't catch and hold after you move it up or down, there is probably dirt in the mechanism. While shade makers don't recommend your doing so, many homeowners solve this problem by prying off the end cap from the flat-pin end of the roller and cleaning the mechanism. Do not, however, under any circumstances attempt to oil the window shade mechanism.

How To Measure, Mount, And Repair Blinds

BLINDS CAN fit either inside the window opening or outside. They generally work better inside, where they are out of the way of the drapes. For an inside installation, measure the exact width and length of the window opening and then consult the dealer for the best stock size.

For an outside mounting, the blinds should extend at least 1½ inches beyond the window opening on both sides. When measuring the length for an outside installation, be sure to allow room

above the opening for the head-box.

In mounting the blinds, you should always use sturdy wall anchors to hold the screws securely in the mounting surface. If you need to repair your blinds, you will find kits available that supply the specific parts—cords, pulls, laddered tapes, etc.—and instructions you need to handle most minor breakdowns.

How To Measure For Shutters

IN MEASURING for interior shutter sets (exterior shutters will block the sun but are generally too much trouble to open and close), it's best to measure the width of the window opening at the top, at the bottom, and at the center since window openings are not always square. Although the shutters can be trimmed to fit, you should use the smallest width measurement when ordering. Do the same thing for length—measuring at the right, left, and center—and then use the shortest height measurement when ordering.

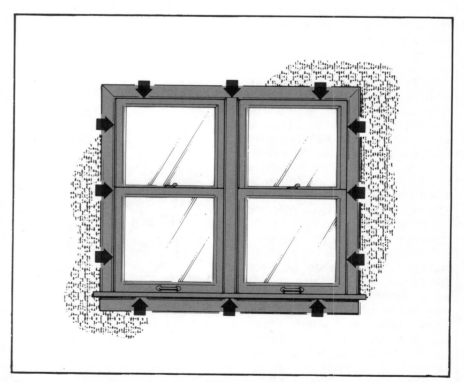

Since window openings are not always square, it's best to measure the width of the window opening at the top, bottom, and center before ordering interior shutter sets. Do the same for the length; measure at the right, left, and center. Then use the shortest width and height figures when ordering.

Shades, Blinds, Shutters, Drapes and Awnings

Materials and Supplies

Shades and Screens

The closer the fit in window shades, the better they do in preventing heat loss. Magic Fit shades from the Newell Companies are made so that you can adjust the fit at home so it is perfect. The shade roller and hemstick are adjustable, and the shade material is constructed with slits at the top so it can be stripped off to fit. The material rips off, leaving a smooth edge.

A way for the homeowner to conserve energy and save additional money by doing-it-themselves is to install insulating shades. A new kit from Madico Consumer Products enables you to make your own shades with Reflecto-Shade, a reflective mesh. The kit includes material, roller, roller end cap for each end of the shade, weights, brackets, nails, a decorative pull-bar and instructions. The material reflects up to 70 percent of the sun's heat rays in summer and helps retain heat in winter. The reflective qualities of Reflecto-Shade are about the same as some reflective films, discussed elsewhere; but because the shades are not actually applied to window glass, they are not as efficient. They can, however, be raised and lowered, an advantage in some cases. (1)

Both heat and glare can be greatly reduced with indoor or outdoor screening for windows such as that made by J.P. Stevens & Co., Inc. The outside application, however, is more efficient. Screening also provides daytime privacy while allowing visibility from the inside. The outside screen, which has a vertical

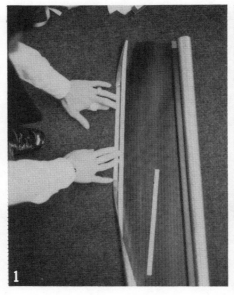 

(1) Reflecto-Shade kit. (2) Comfort Shade indoor screen.

weave, is called Comfort Screen. The indoor screen, with a horizontal weave, is called Comfort Shade. The fiberglass screening can be installed in frames, rollers or tension systems. It carries a 10-year guarantee against corroding or denting. Comfort Screens are available in white, gray, charcoal and bronze; Comfort Shade is only available in white. (2)

While storm windows do a good job in conserving your home's heat in winter, they are not too effective in providing relief from excessive solar heat coming through your window in summer. The hot air trapped in the cavity between the storm window and the prime window glass can be radiated inside your

rooms even after the sun has set. To avoid this, you should try to block as much heat outside the window as possible. One way is to use a product like Solar Shields made by Vimco Corporation. These are specially-woven fiberglass screens stretched taut with aluminum rails that attach to the head and sill of a window. The screen mesh can stop about 75 percent of the sun's heat outside the window.

A louvered aluminum window screen developed by Kaiser Aluminum is able to reduce solar heat gain through windows by as much as 80 percent, resulting in substantial savings on air conditioning requirements and energy costs. The solar control product, called ShadeScreen, consists of tiny built-in aluminum louvers that intercept direct sunlight but allow controlled useful light to enter. Panels can be removed and stored for winter months, allowing more sun to enter rooms and reducing heating costs. (3)

Shutters

It is recommended that you stop as much cold and wind in winter and heat in summer as possible before they get into your home. One good way to accomplish this is with exterior shutters. Exterior rolling shutters from the Swiss Blinds Division of the Pease Co. insulate all year round and act as storm windows

due to the double glazing effect. The shutters are operated from inside the home and can be opened for partial light control and closed for privacy and security. (4)

Besides blocking out the sun when necessary, shutters—such as the Ever-Strait Rolling Shutters made by the Pease Co.—add insulation because each slat is an extruded double wall with dead air space. The product also adds to your home security since they can be locked in the down position. The shutters are available in a variety of widths to fit standard window openings. For patio doors and other wide openings, two or more rolling shutters can be installed side by side. (5)

Awnings

If you are unable to obtain awnings in your area, you can order a number of different styles and sizes from Sears and Montgomery Ward. Both offer stationary and movable aluminum awnings. They also come in several colors or two-tone combinations. Most styles have matching door canopies.

Awnings and canopies on your home will reduce the load your air conditioning system has to bear in the summertime. A line of aluminum awnings and canopies, called Sunshields, is available from the Kinkead Divi-

60

sion of United States Gypsum. The reinforced units are finished with an acrylic coating that will not peel, chip or bubble. Door canopies come in a variety of widths; the awning line includes adjustable roll-up awnings and four position fold-down awnings, one with a complete window cover. (8)

Awnings made of canvas are available to shade room air conditioners as well as all windows throughout a house. Other energy-saving uses include windbreaks, carport protectors and patio covers. A patio cover, for example, can shield the concrete from sunlight that is reflected and radiated into your home. A movable cover allows you to use

this extra heat during the winter. Canvas comes in a variety of weights, finishes and weaves. And, canvas awnings can be fashioned in many different ways to fit the architecture of your home. The material also is sold as yard goods so you can cover existing frames with new material. (7)

The starting points for cooling your home efficiently are shade and insulation, whether or not you have air conditioning. Vented awnings such as those made by the Howmet Aluminum Corp., not only do the shading an awning should do, but also allow air to go through so there are no hot air pockets. The firm also offers

other types of awnings, both movable and fixed. Howmet awnings, both ventilated and others, are available in all-aluminum construction or in rigid vinyl. They come in a wide variety of styles and color combinations. Both aluminum and vinyl awnings have a 10-year guarantee.

The sun should not shine directly on your window air conditioner. To allow your air conditioner to work less and lower your utility bills, the unit should be shaded. Awnings are one answer, and Artcraft Industries, a division of Enamel Products & Plating Co., makes aluminum awnings with a baked-on white enamel finish. They come in two widths—36 and 48 inches . (6)

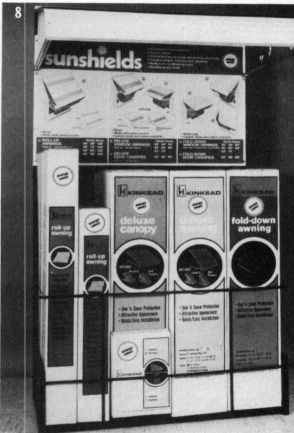

(3) ShadeScreen. (4) Swiss Blinds exterior rolling shutters. (5) Ever-Strait Rolling Shutters. (6) Artcraft aluminum awning. (7) Canvas awnings. (8) Sunshields awnings and canopies.

61

Heating

YOU CAN—and should —button up your house through insulation, caulking, weatherstripping, and other such means, but unless your heating units are in top working condition, you will continue to waste sorely needed and increasingly expensive energy. Efficiency is the key concept here, and even if you don't understand all the technical aspects of your home's heating system, here are several steps you can take to enhance your system's efficiency—and thereby conserve fuel.

• Keep your heating system—i.e., filters, blower fans, ducts, vents, and even the thermostat—clean. No matter what type of heating system you have in your home, it will operate at peak efficiency only when it is free of dirt.

• Protect your thermostat from anything that would cause the instrument to render a false reading. If the thermostat is in a draft, misplaced on a cold outside wall, or too near a heat-producing appliance (such as a TV or lamp), its accuracy in maintaining a comfortable temperature in your home will be compromised.

• If you will not be at home for a few days, turn the thermostat to its lowest setting. And if there is no danger of pipes freezing during that interval, turn the heating system off completely.

• Avoid constant thermostat adjustments; they waste fuel in most cases. When you first come into the house after the thermostat has been turned down, don't set it higher than the desired temperature. Setting the thermostat up very high generally will not cause the temperature to reach the desired level any faster, and you could forget to dial down and thereby overheat your home.

• On the other hand, one adjustment you should make is a reduction in the thermostat setting before you go to bed every night. Cutting back for several hours can make a big difference in fuel consumption.

• You can also reduce the thermostat setting while preparing meals. Heat generated by cooking will compensate in most cases.

• Similarly, reduce the thermostat setting when you have several people in your home. People generate heat, and a large group can quickly make your home uncomfortably warm.

• Close the drapes over large windows and glass doors to form a barrier against heat loss.

• If there are rooms in your home that you seldom or never use, close the vents in these rooms and shut the doors that lead from these rooms to the occupied portions of your home.

• Keep the damper in your fireplace closed except when you have a fire going. Otherwise, updrafts will suck out your home's heated air through the chimney.

• Maintain proper humidity in your house. A house that is too dry can feel uncomfortably cold even when the temperature setting is correct.

• Never allow furniture or drapes to block heating vents, and don't build covers or shelves to hide radiators. All such furnishings can lock in the heat that is supposed to circulate throughout the room.

• If you have children, consider installing automatic closers to prevent exterior doors from being left open unnecessarily.

• Allow hot bath water to cool before draining the tub. The heat from the water will be picked up by the air and help keep your home comfortable.

How Heat Travels Through Your House

WHETHER YOU have a furnace in the basement or individual space heaters in every room, the warmed air must circulate throughout the house or room or else you'll only be warm when hovering over the unit. Home heating systems conduct warm air in several ways.

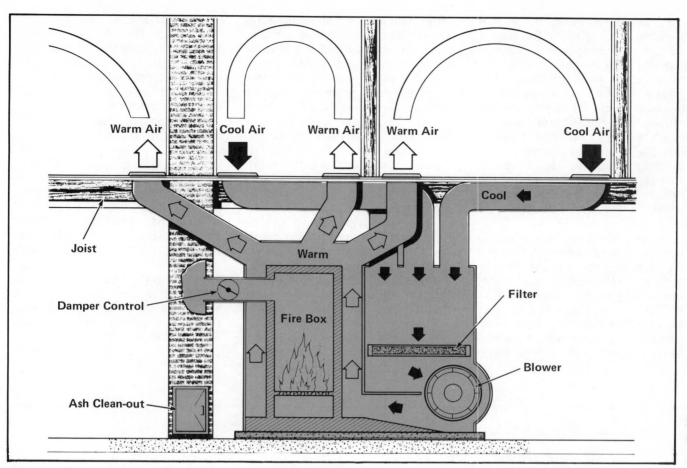

A forced air heating system employs a fan to move the warmed air through ducts to rooms far beyond the furnace.

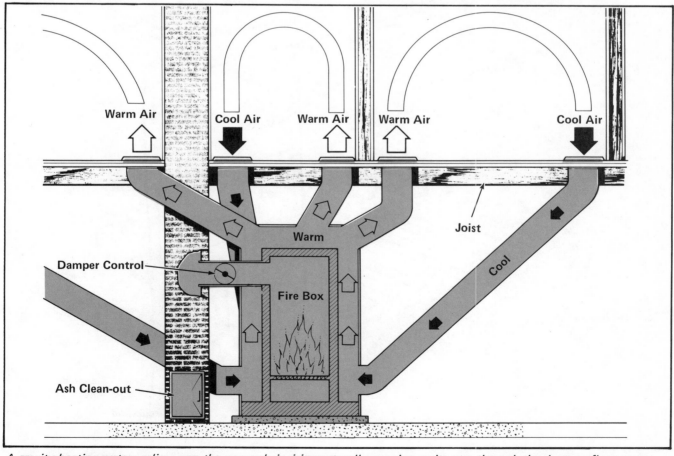

A gravity heating system relies upon the warmed air rising naturally, moving on its own through the ducts to floor vents.

Forced Air Heating. Because it employs a fan to move the warmed air, a forced air system is probably the most effective way to conduct heat through your house. The fan can move the air through ducts to rooms far removed from the furnace, and the same fan can also bring cold air into the heater for warming. In a forced air system, consequently, air keeps moving throughout the house. These systems can be fueled by gas, oil, or electricity.

Gravity Heating. The principle behind gravity heating is that hot air rises because it is lighter than cold air. Of course, the heater unit must be down near or below the floor for this type of system to be effective, and the best location for a central gravity flow system, naturally, is in the base-

ment. The warmed air rises and is carried through ducts to vents in the floor throughout the house. On the other hand, if such a central unit is located on the main floor, the heat registers must be high up on walls since they must always be higher than the furnace. Once in a room, the warmed air rises toward the ceiling; as it cools, it comes back down, enters the return air ducts, and goes back to the furnace to be reheated.

The big advantage to gravity heating is that the furnace needs no motor. Therefore, it uses no electric current and its air movement is silent. Gravity heated air, though, does not move with much force and thus is usually not able to go through a filter; nor does the system work well if the heated air must travel

long distances. Slow movement, moreover, allows for greater heat loss before the air ever gets to the room. Finally, gravity heating systems cannot warm a house as evenly as do most forced air systems.

Wall heaters can be of the gravity type, having a return air vent at the bottom and a vent for the hot air to go out at the top. Floor furnaces can also operate on the gravity system.

Radiant Heating. Radiant heating systems function by warming walls, floors, or ceilings, which then warm the air with which they come in contact. Usually the heat source is hot water that is circulated through pipes embedded in the wall, floor, or ceiling.

Radiant heating is often built

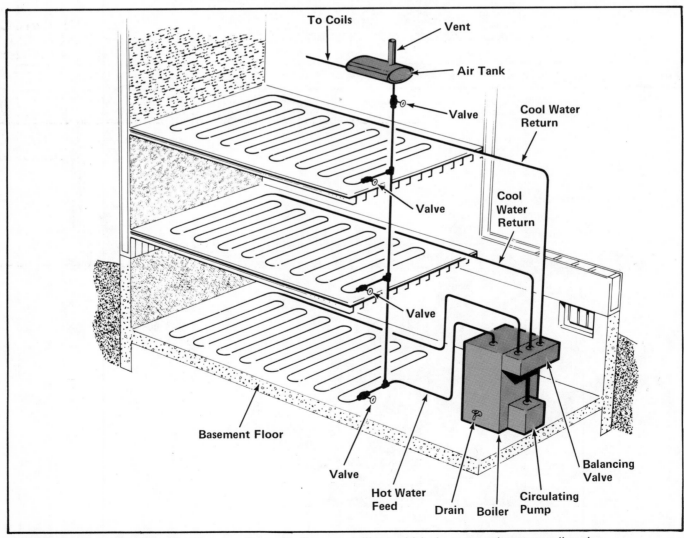

A radiant heating system functions by warming walls, floors or ceilings, which then warm the surrounding air.

into houses constructed on a slab foundation. A network of hot water pipes is laid within—but near the surface of—the slab. When the concrete is warmed by the pipes, it warms all the air that contacts the floor surface. The slab need not get very hot due to the large area involved; it will eventually contact all the air throughout the house.

Caring For Your Central Heating System

NO MATTER what fuel your central heating system consumes, there are some routine steps you can take (without any extensive training, experience, or tools) to keep your system functioning at peak efficiency.

The worst enemy of your central heating system is dirt. Dirt can severely degrade heating efficiency and thus waste fuel. Furnace filters, of course, are present in the system to catch dirt, preventing it from fouling the furnace or the air supply. Most homeowners realize that furnace filters must be cleaned or changed periodically, but most rely on the preseason inspection by a serviceman for filter maintenance.

Once or twice a year, however, is not enough filter care. In fact, many central heating systems need a filter change once a month. If your heating unit has a permanent filter, clean it according to the instructions given on the heating unit itself, and then spray the filter with a specially formulated filter coating chemical (available at hardware stores).

Most central heating systems, fortunately, have disposable filters that are quite inexpensive and easy to change. The filter is usually located between the blower and the return air vent. To replace the filter just follow these simple steps.

Step 1. Remove the filter access panel on the front of the furnace.

Step 2. Remove the old filter. It is either held in place by some sort of a removable device or else it is in a slot. The size of the filter should be on its sides or ends.

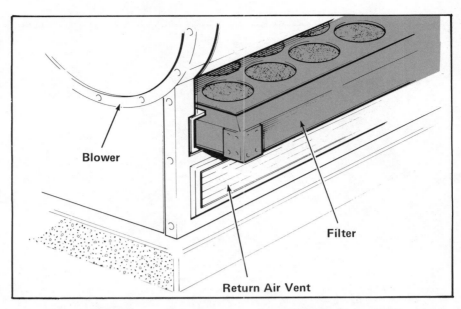

After removing the filter access panel on the front of the furnace, take out the old filter; it is either held in place by some sort of a lift-off device or else it is in a slot. Be sure to install a new filter of the correct size, and check that the arrow on the filter points in the same direction the air flows.

Step 3. Install a new filter of the correct size. Make sure that the arrow on the filter points in the same direction that the air flows.

Step 4. Replace access panel.

What most homeowners fail to realize is that there are other places inside a heating unit where dirt can disrupt normal operating efficiency. The blower or fan that moves air from the unit out through the ducts can get dirty, too. This is particularly true with a squirrel-cage type fan. When the cage openings get clogged, the fan cannot move enough air, and the system runs inefficiently. Here are some blower cleaning instructions.

Step 1. Turn off all electric power to the unit by disconnecting the appropriate fuse or tripping the circuit breaker switch. Make sure that the current is off before you start working on the blower.

Step 2. Remove the filter access panel. The fan is usually held in a track by sheet metal screws. Remove the screws.

Step 3. Slide out the fan unit. If the electric cord leading to the fan unit is not long enough for you to slide the unit all the way out, disconnect the cord after

noting carefully exactly how the wires are hooked up.

Step 4. You must brush off each fan blade. An old toothbrush may be the biggest brush you can get into the spaces between the blades.

Step 5. Clean the entire area around the fan. Here you can use a larger brush.

Step 6. Go over the fan unit with the hose from a canister vacuum cleaner to remove all the dirt you loosened.

Step 7. Replace the fan assembly, securing it with the screws you removed earlier.

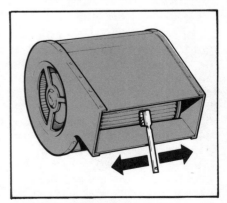

Clean each blade of the furnace blower fan with an old toothbrush.

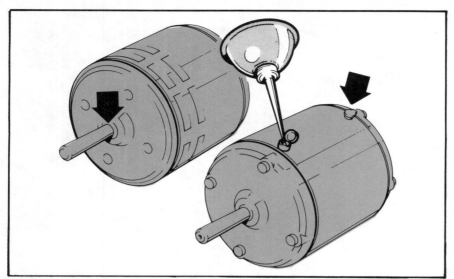

Apply lubricant to the fan motor oiler cups (right) at the start of every heating season. With a sealed unit (left), you need only squirt some oil on the felt pad where the shaft enters the motor.

Make sure that the vents and return air registers are clean. A canister vacuum cleaner does a good job of removing any lint from them, but in some kitchen areas airborne grease can also collect on the vents, and you must remove and wash such grills to get them clean.

Some parts of your heater unit require lubrication as well as cleaning. Be sure to cut off the power to the unit before reaching in to lube. Here are the spots to lubricate.

1. The blower motor has oiler cups. Lift the cups and squirt in from five to ten drops of lightweight oil at the beginning of each heating season.

2. If the blower motor is belt driven, spray belt dressing on the belt once a year and any time you hear the belt squealing.

3. Apply lightweight oil to the fan motor in the condensing unit at the start of each season. If the fan blade is aimed upward, look for a rubber or plastic cover over the oiler cup; lift the cover to oil, but put in no more than five to ten drops. If the fan faces out, look for oiler cups on the motor housing; to reach them, you must remove the access plate. If there are no oiler cups, you may have a sealed unit which requires no oiling. Since these units do

have a felt pad where the shaft enters the motor, however, it is a good idea to squirt oil on the pad about every four years to prevent it from drying out.

These simple care and maintenance steps will make your unit work more efficiently and thus reduce your utility bills. They should also cut down on those expensive repair bills you pay every year.

Gas Furnaces

ALTHOUGH MORE expensive than fuel oil in many parts of the country, natural gas is generally the more popular and preferred source for heat. Natural gas burns cleaner than fuel oil, and most natural gas furnaces present far fewer operational difficulties. In fact, the problems that do afflict natural gas furnaces usually have little to do with the fuel source itself; instead, they typically involve the

thermocouple, the pilot, or some aspect of the electrical hookup. And despite the fact that natural gas costs more (and its price is rising steadily), the natural gas furnace is less costly both to buy and to install than its fuel oil counterpart.

The basic bit of maintenance that you can perform on your gas burner is to change or clean the filters at the prescribed periods. If some serious difficulty afflicts your furnace, by all means call in a professional repairman. And any time that you smell gas or suspect a gas leak, close all valves and call the gas company immediately.

There are, nonetheless, some simple repairs which you can perform yourself and save the cost of an unnecessary service call.

1. If the pilot goes out, check the orifice to see whether it is clogged; if it is, clean it out with the point of a needle.

A faulty pilot may well signal a problem with the thermocouple, however. You can replace a thermocouple which is no longer functioning properly, but before you replace it check to make sure that the pilot flame is bathing it satisfactorily. You can adjust the pilot flame to bathe the tip of the thermocouple by following the instructions in the owner's manual for your furnace.

2. If the burner itself fails to light, the first thing to check, of course, is the pilot. If the problem is not the pilot, it may be a faulty gas valve that is sticking when it shouldn't. Tap the gas valve sharply with your knuckles.

Another possible cause is a faulty thermostat. Turn the thermostat to a higher setting; if the gas burner goes on only when the thermostat is set much higher than it should be, replace the thermostat.

The only other possible cause for the burner not functioning that you should investigate before calling a repairman is that no electrical current is reaching the furnace. Check for a blown fuse or tripped circuit breaker switch and replace or reset as necessary.

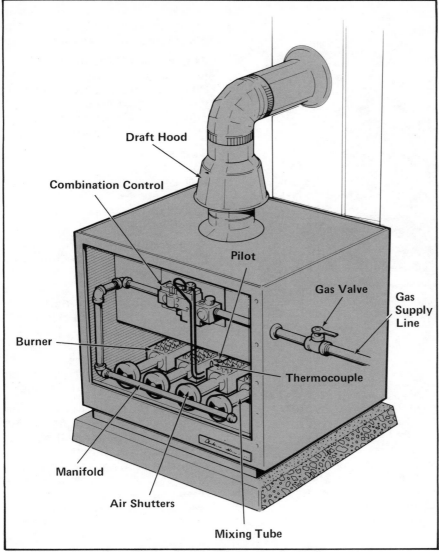

Draft Hood

Combination Control

Pilot

Gas Valve

Gas Supply Line

Burner

Thermocouple

Manifold

Air Shutters

Mixing Tube

Most natural gas furnaces present far fewer operational difficulties than do oil-fired furnaces. Problems usually involve just the thermocouple, pilot light, or some aspect of the electrical hookup.

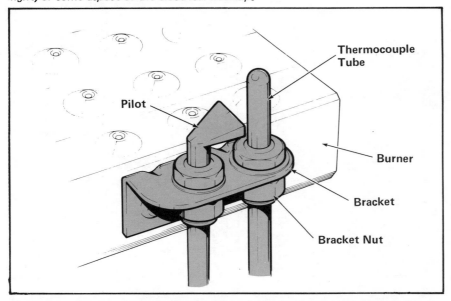

Thermocouple Tube

Pilot

Burner

Bracket

Bracket Nut

Check to make sure that the pilot light is bathing the thermocouple properly. You can adjust the flame to bathe the tip of the thermocouple tube.

Oil Fired Furnaces

OIL-FIRED BURNERS are used in many sections of the country as the basic heat source for a warm air heating system, a hot water heating system, or a steam heating system. Most of the home oil systems in use today are called pressure burners. A high-pressure spray combines with a blower to send a fine mist of oil into the combustion chamber, where the oil is ignited by an electric spark. The oil continues to burn, of course, as the mist continues to be sprayed.

Although oil burners are generally reliable, hard-working systems that provide years of trouble-free service, you must know how to maintain them properly. Here are the steps in maintaining an oil burner.

Step 1. Keep the fan motor lubricated according to the instructions in your owner's manual. Most motors require the application of an electric motor oil.

Step 2. Keep the blower clean. Again, follow what the manual says, but be sure to shut off the current before you start cleaning the blower. A small skinny brush usually works best.

Step 3. Keep the stacks free of soot deposits by dismantling the stacks and bringing each section down sharply against the newspaper-covered floor. Be sure to seal the stack connections properly when rejoining the sections.

Step 4. Check for leaks around the mounting plate by passing a lighted candle slowly around its rim while the burner is going. The slightest leak will suck the flame toward the plate. Usually tightening the bolts stops the leak, but you may need to replace the gasket. Often, refractory cement applied around the rim stops any leak.

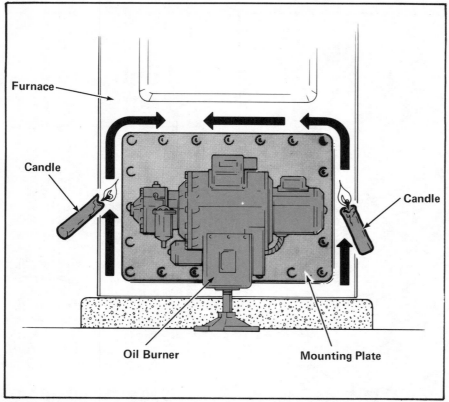

Check for leaks around the mounting plate of an oil burner by passing a lighted candle slowly around its rim; a leak will suck the flame toward the plate.

stack control. If you see a heavy coat of soot, however, it is a sign of incomplete combustion, and you should call for service.

Step 6. Inspect the air draft regulator. It should close when the unit goes off and be slightly open when the burner is on. You can adjust the air draft regulator easily by turning the counterweight.

Step 7. Consult the owner's manual for your furnace regarding the proper adjustment of the spark gap, the air tube shutter, and all other components requiring adjustment.

Step 8. Call in a pro to clean, adjust and inspect the unit prior to the onset of winter.

If the unit fails to run, here are some of the steps you can take to get it going again.

Step 1. Check to be sure the switch is on; if it is, push the reset button if the unit has one.

Step 2. If nothing happens, check the fuse or circuit breaker switch to see whether the power is still on. If you find that the fuse or circuit breaker goes out again after you put in a new fuse or reset the switch, you know that your system has a short. Repair it or have it repaired immediately.

Step 5. About two months into the heating season, clean and inspect the heat sensor in the stack control. Cut off current to the furnace at the electrical entry box, remove the mounting screws, and slide out the stack control to expose the sensor. A light coating of soot is normal; just clean the heat sensor and replace the

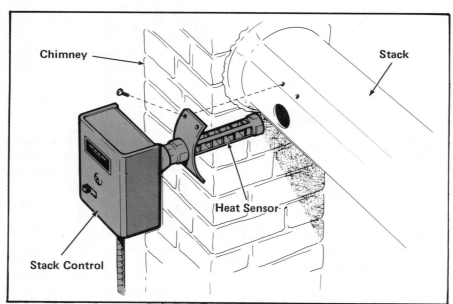

Clean and inspect the heat sensor in the oil burner's stack control. After cutting off the current to the furnace, remove the mounting screws and slide out the stack control to expose the sensor. Clean and replace it.

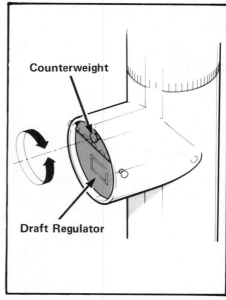

Adjust the air draft regulator by turning the counterweight so that it is open slightly when the unit is on.

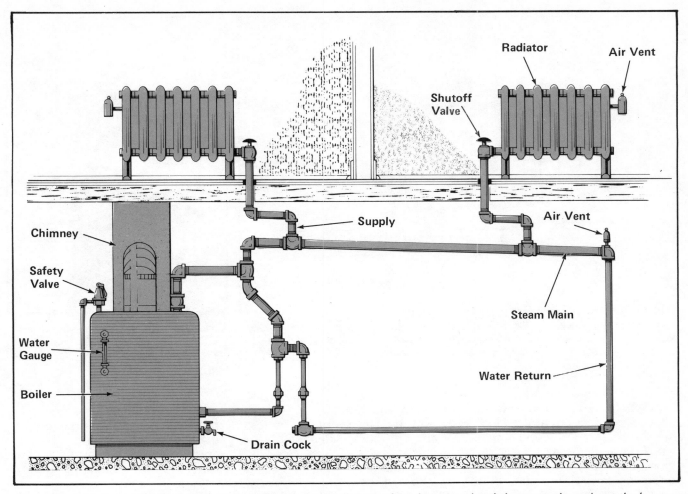

Steam heat rises naturally from the boiler, passing up into radiators, where it warms the air in rooms throughout the house.

Step 3. If the burner works spasmodically or sputters before going out completely, you probably have a dirty oil filter. Check your owner's manual for instructions on getting to the filter and on bathing the filter in kerosene. Since kerosene is highly flammable, however, you must exercise extreme caution when cleaning the oil filter.

Step 4. Make sure you have enough fuel oil.

Step 5. Move the thermostat up to a higher than normal setting to check its accuracy. If the burner goes on only when the thermostat is set much higher than it should be, replace the thermostat.

Step 6. If you have a steam system, be sure the water level in the boiler is adequate.

Step 7. If you have a hot water system, check the recirculating pump to make certain that it is moving the water around as it should. If the water does not circulate, the burner system does not operate.

Steam Heat

STEAM HEAT is not installed in homes anymore, but it is such a durable heating system that many homes and apartment buildings are still heated by steam.

Basically, a steam heat system works in the following manner. There is a boiler in the basement which heats water until it turns into steam. The steam rises naturally, going up to the radiators and warming the air in rooms throughout the house. As the steam cools it liquefies, and the water then flows back down the same pipes to the boiler where it is heated once again.

It is a simple system, but for it to work properly pipes and radiators must slope back toward the boiler. If the water cannot run back to the boiler, it collects and soon blocks the path of the steam. This is usually accompanied by hammering noises in the system, and the blockage may lead to one or more individual radiators failing to function.

If you note such noises or malfunctions, you can probably correct the cause of the problem easily. Merely place blocks of wood under the affected radiator(s) to correct the angle of slope. If you suspect that the

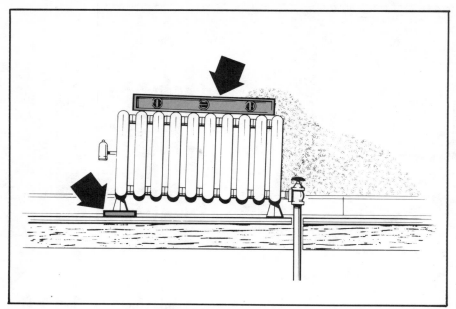 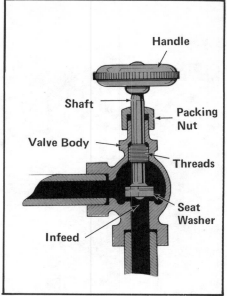

For a steam heat system to function properly, all pipes and radiators must slope back toward the boiler. Check the angle of slope with a level, and correct incorrectly tilted components to restore proper operation.

Each individual radiator valve must be opened all the way for the radiator it controls to function correctly.

pipes which carry steam upward and water downward are at fault, check their angle of slope with a level. Frequently, these pipes become incorrectly tilted when the house settles, but you can solve the problem by applying pipe straps to reestablish the proper slope.

Other problems can afflict steam heating systems, and although these problems occur less frequently than the radiators or pipes losing their correct angle of slope, they are problems you should be able to recognize and repair.

1. If there is no heat throughout the system, then either the burner is out or the water level is too low. Follow the directions appropriate to your heating system (gas or oil) to get the burner going again. The gauge on the boiler will tell you whether the water level is correct; if it is too low, the system will shut off automatically. Check the water level gauge periodically and add water as needed through the fill valve.

2. If an individual radiator is cold and both it and the pipes leading to it are tilted properly, check the inlet valve. The valve must be opened all the way for the radiator to function correctly.

3. When you notice that the entire system is operating at less than optimum efficiency, the problem is most likely clogging due to rust and scale. If you see any rust in the water level gauge tube, shut off the boiler and allow it to cool. Then flush the system by draining and refilling it several times. Finally, add one of the commercially prepared products formulated to curb the buildup of rust and scale in boilers.

4. If you detect more heat going to some radiators than to the others, it probably means that the vents are not adjusted properly. Adjust the vents so that the ones farthest from the boiler are opened more than the ones closest to the boiler.

5. Leaks around inlet valves, radiators, and pipes are plumbing problems that you may be able to correct or which may require the services of a professional. There are special additives which you can put into the boiler's water supply; these additives are formulated to stop leaks in steam radiators. Pipe leaks are frequently due to loose connections which can generally be tightened with a wrench, but leaks around inlet valves usually signal that the stem packing is no longer doing

its job and must be replaced. To replace the packing, first shut off the boiler and allow the system to cool. Then proceed as you would in replacing the packing around a faucet handle.

Hot Water Heat

SINCE WATER retains heat for some time, it can be used effectively in home heating systems. Such systems are either of the gravity type or the hydronic—i.e., forced hot water—type, with both types of hot water heating systems capable of being fueled by gas, oil, or electricity.

The gravity system depends on hot water rising to circulate heated water from the boiler through a system of pipes. The pipes are nothing more than plumbing pipes, which take the hot water to a radiator unit in each room. The heat from the water is transferred first to the metal radiator and then to the air via convection currents. As the water loses its heat, it becomes heavier and starts to flow on back to the boiler through the return pipes. Most gravity systems heat the water to no more than about 180 degrees, and the cooled water that goes back to the boiler rarely falls below 120 degrees.

There are two types of gravity systems—open and closed. The open version has an overflow outlet to let water escape, preventing a buildup of excessive pressure in the system. The closed system has a sealed expansion tank. When the water pressure builds up in the system, the excess water goes into the expansion tank to prevent ruptured pipes or boiler.

The hydronic hot water system is much like the closed gravity system except that it employs a motor-driven circulating pump (called a circulator) to move the water. As a result, the water in a hydronic system moves more rapidly and arrives at the room radiator with less heat loss.

The better radiators for hot water systems are called convectors. These units employ a series of fins to maximize the surface area for air contact and heat transfer.

Here are some maintenance tips for keeping a hot water heating system at peak operating efficiency.

1. Boiler problems are usually caused by lime deposits in the water which form scale inside the pipes. Cleaning compounds are available that homeowners can introduce safely into the system to dissolve the scale.

2. The circulator in a hydronic system must be kept in good operating condition. The pump motor should be cleaned and lubricated regularly.

3. If air gets trapped in a convector radiator, the hot water may not be able to enter. The unwanted air can be purged by turning an air vent valve on the unit until the hissing stops and water comes out. Many valves accomplish this air release automatically.

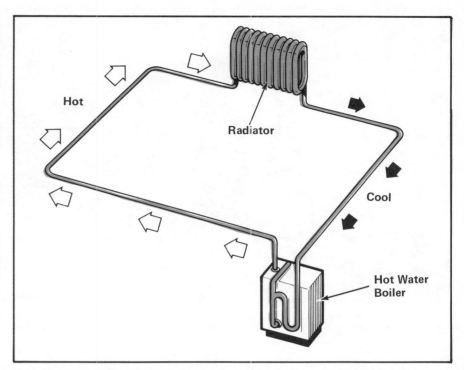

With hot water heat, the heated water leaves the boiler, travels to the metal radiator where it loses its heat, and then returns to the boiler.

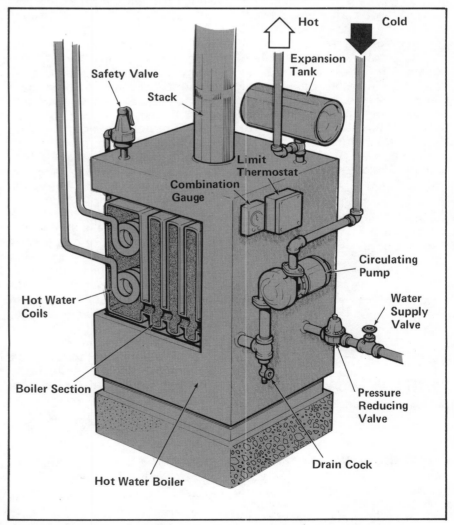

Hydronic hot water heating systems employ a pump to move the water.

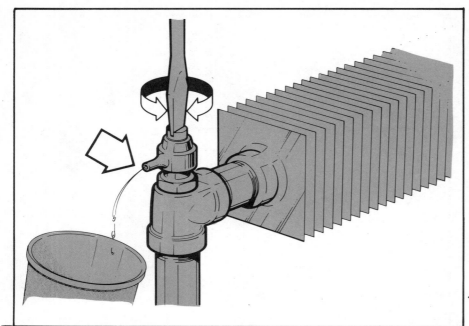

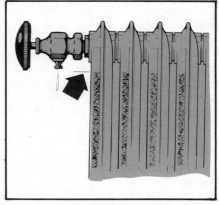

▲ Purge unwanted air from a radiator by turning the valve until the hissing stops and water comes out.

◄ The better radiators for hot water systems are called convectors. They employ a series of fins to transfer heat.

4. Air in the expansion tank can also be a problem. Too much air or too little air affects the pressure in the tank. Many systems are set up to add water or drain it off automatically to maintain constant air pressure; others require the manual opening of valves to release excess pressure or the pumping of air into the tank to increase the pressure. Consult your owner's manual for the appropriate instructions for your particular system.

5. Some hot water systems require yearly flushing during the summer. Consult your owner's manual or a competent heating firm in your area.

Electric Heating Systems

ELECTRICAL HEATING is expensive. This is true whether you have a central warm air furnace, a boiler system, or baseboard units to heat individual rooms. While an electric heating system has its advantages, its operating cost generally makes it less desirable than the other systems available. The high cost also means that minimizing heat loss through improperly installed ducts or through inadequate insulation is even more important with electrical heating than with other heating systems.

On the positive side, since no combustion takes place in an electric heating system, it is a much cleaner system than the fuel-burning types. Moreover, since no flue is required to carry off undesirable combustion materials, no heat is lost through such flues as it is in gas and oil systems. Finally, the only moving parts in an electric heating system are the blower fan or circulating pump, making the system easy to install and very easy to maintain.

In fact, there is little you can do to maintain an electric warm air furnace except keep it clean. To do that, follow these simple steps.

Step 1. Clean or replace the filter frequently.

Step 2. Keep the blower fan clean and properly lubricated.

Step 3. Clean all vents and registers periodically.

Step 4. It is also a good idea to check the ducts for signs of heat loss due to loose joints or damaged or faulty insulation.

Almost any problem in the system is bound to be electrical in nature. If the unit fails to heat, therefore, perform the following checks before calling a professional in to service the unit.

1. Check to be sure that there is no interruption in the flow of power to the unit. Look for a tripped switch at the circuit breaker box.

2. Each heating element within the furnace is protected by its own fuse or circuit breaker. Shut off the current to the unit and check for blown fuses or tripped circuit breaker switches.

3. Check to be sure that the thermostat is properly set and that it is functioning as it should.

4. If these checks do not reveal the trouble, then the next step would be for a professional to troubleshoot the wiring and the heating elements, looking for a break in the circuit.

Another type of electrical heating device is the individual baseboard unit. An ideal way to supplement a central heating system in a particular room, the

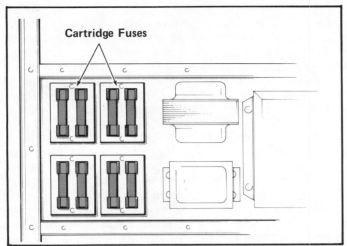

Each heating element within an electric furnace is protected by its own fuse or circuit breaker. Check for blown fuses or tripped circuit breaker switches.

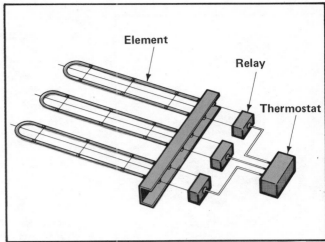

If you cannot detect any fault with the thermostat, fuses or circuit breakers, have a professional inspect the wiring and heating elements for a circuit break.

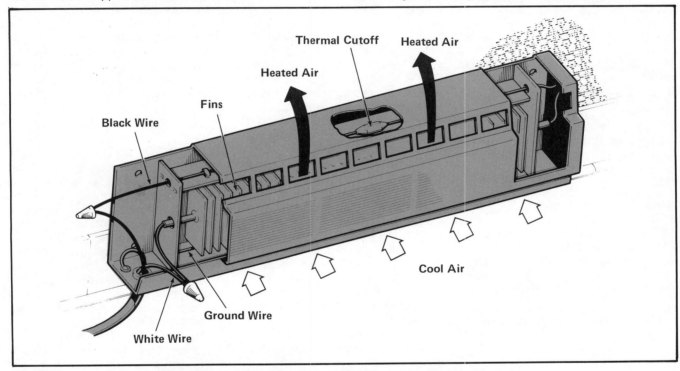

Most baseboard heaters send out heat via convection; there is seldom a fan to malfunction.

baseboard heater is also quite popular for heating a new room that has been added onto the basic structure. Since it hugs the wall, the baseboard heater takes up very little floor space. And since the baseboard unit involves no plumbing or ducting, it can be installed easily; some types just plug into a wall outlet. Heat is usually sent into the room by convection, which means that there is no fan to malfunction, and each unit has its own thermostat (usually built into the

heater), permitting individual adjustment of the desired temperature for each room in which a baseboard unit is located.

Like a central electric heating system, the baseboard unit requires no routine maintenance tasks, and nearly all problems are electrical in nature. Here are the basic remedial steps to take if you discover that no heat is coming out of your electric baseboard heater.

Step 1. Check for a blown fuse or

tripped circuit breaker switch. Replace the fuse or reset the circuit breaker switch.

Step 2. Shut off the current and then carefully inspect the unit's wiring; check both by looking at the wiring and by testing it with a continuity tester.

Step 3. Remove any obstruction that could be impeding the air flow. Something may have fallen into the unit, or the drapes or a piece of furniture may be blocking the air flow. Once you remove

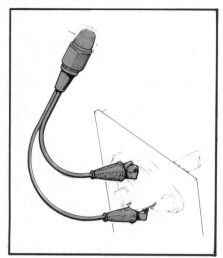

Check the thermal cutoff safety device with a continuity tester. Replace the device if you find it defective.

the obstructions, push the unit's reset button.

Step 4. Check the heating element with a continuity tester. If it is defective, replace it according to the instructions in your owner's manual or have a professional repairman replace it for you.

Step 5. Check the thermal cutoff safety device with the continuity tester. This device shuts the unit off when a problem exists. If it is faulty, however, it will prevent the unit from running. Replace cutoff device if it proves defective.

Step 6. Check the thermostat and replace it if you find it defective.

The Heat Pump

MANY PEOPLE think that the latest thing in home heating and cooling is the heat pump. Actually, though, the heat pump first appeared about 40 years ago. Unfortunately, the performance of the early models—sometimes referred to as reverse cycle air conditioners—suffered due to the lack of existing knowledge about the action of refrigerants under the reversed cycle, the lack of knowledge about how sealed parts could cope with the stresses to which they were subjected, and the lack of trained service people in the field. These deficiencies resulted in the early heat pumps failing to make the grade in the consumer marketplace.

Most of the heat pump's problems have now been solved. Better materials are available, and manufacturers are creating more reliable equipment. For example, where the old-fashioned reverse cycle air conditioners were effective only in milder climates, the new heat pumps can handle really cold weather as well (although some homes may require supplemental heating systems). On the other hand, in areas where air conditioning during the spring and summer months is not necessary, the heat pump offers few advantages over conventional heating systems.

In deciding whether you should install a heat pump, you must do some homework. First determine how much the installation will cost. The heat pump itself costs more than do the other types of central heating and air conditioning systems. Get two or three estimates. Next, ask your local power company for a cost estimate on operating the heat pump, and then compare this estimate with the operating costs of other types of heating and cooling systems. If a heat pump can save you money in terms of operating costs, then proceed to calculate how many years it will take for the heat pump to pay back its higher installation cost. Divide the estimated annual operating savings into the additional installation cost, and then decide whether the heat pump is worth the extra expenditure. Keep in mind that gas and oil prices are going to increase significantly during the years ahead.

Another factor to consider is who will install your heat pump. Having a contractor who can determine the size unit your home requires and who can then install your heat pump properly is crucial to your buying decision. If there is no one in your area who is experienced in heat pump installations, then you had better stick with the more conventional heating and cooling systems for at least the present time.

Heat pump maintenance is also important. Small problems which are not properly and promptly rectified can result in very expensive compressor problems. Since maintaining a heat pump is more technical than caring for the average heating system, make sure that trained service people are available. About all the average homeowner can do is to keep the system free of dirt—i.e., keep the filter clean and remove any other obstacles to the flow of air. Your owner's manual will provide details on cleaning and possibly on other maintenance steps you can take.

How A Heat Pump Works

THE HEAT PUMP does not burn fuel to produce heat, nor does the electricity it consumes go through an element. It functions, instead, according to a principle long known in refrigeration mechanics: A liquid absorbs heat as it turns into a gas and releases heat as it returns to a liquid state.

During the heat pump's cooling cycle, the liquid refrigerant is pumped through a coil of tubing. The liquid expands as it moves through the coil, turning into a gas and absorbing heat from the air surrounding the coil. A blower fan then pushes air around the cooled coils through ducts and into the house. The gas—now carrying considerable heat—moves through a compressor which begins the liquefying process and then to a condenser coil outside the house where the compressed gas releases its heat and returns to a liquid state.

During the heat pump's heating cycle, the process is simply reversed. The liquid refrigerant is pumped the other way—toward the coil outside the house. Here it picks up heat from the outside air as it turns into a gas. The gas is then compressed and fed to the coil inside the house where it re-

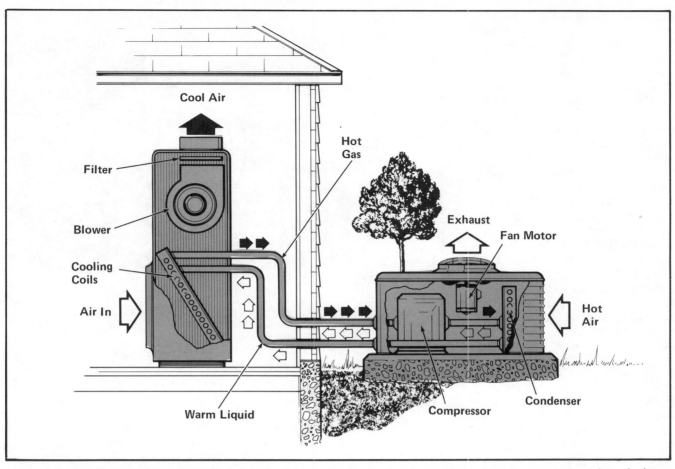

In summer, the heat pump functions as an air conditioner, blowing air over the cooled coils and through ducts into the house.

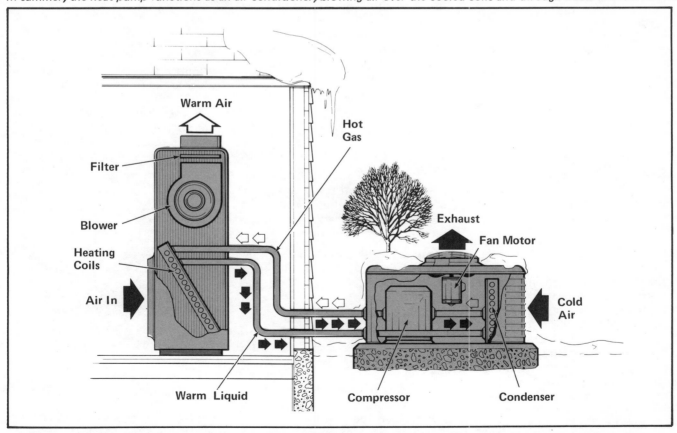

In winter, the heat pump picks up heat from the outside air and feeds it inside for distribution throughout the house.

leases its heat as it again turns to liquid. This heat is then blown through the ducts and into the house.

When the outside temperature is very cold, an electrical heater comes on to supplement the heat pump's output. Like standard electric heating systems, this supplemental unit is quite expensive to operate. The heat pump's advantage over other electric heating systems, of course, is that it extracts the free heat from the air. Even cold air has some heat.

The heat pump is an extremely efficient heating mechanism when the temperature is around 45 to 50 degrees Fahrenheit. For every BTU it consumes it produces from two to three BTU's. As the temperature goes down, however, the heat pump becomes less efficient.

The heat pump is a fine system right now and will certainly be improved in the years to come. At present, it cannot handle extremely cold situations without a supplemental backup heating system of more conventional design. Nevertheless, the heat pump certainly merits the consideration of anyone either building a home or replacing an existing heating system.

Thermostats

THE THERMOSTAT that controls your home's heating system is a highly sensitive instrument, responding to even the slightest changes in temperature. If it is incorrectly calibrated, however, the thermostat can be wasting a good deal of energy. Suppose you set your thermostat for 70 degrees, but the actual temperature in your house is 73 degrees. Those extra three degrees of heat—which you neither need nor want—cost you money.

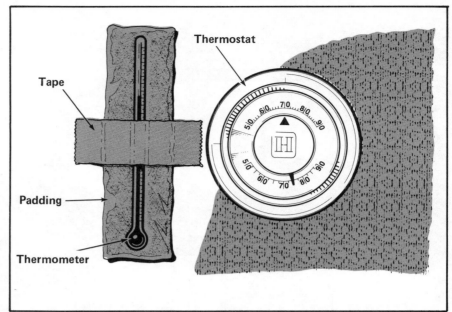

To check the accuracy of a thermostat, tape a glass tube thermometer to the wall a few inches away and compare the readings.

Here's how to check your thermostat.

Step 1. Tape a glass tube thermometer to the wall a few inches away from the thermostat. Place a small tab of padding under the thermometer to prevent it from actually touching the wall, and make sure that neither it nor the thermostat is subject to any outside temperature influences. Often the hole in the wall behind the thermostat through which the wires come is too large, allowing cold air to reach the thermostat and affect its reading.

Step 2. Wait about 15 minutes for the thermometer to stabilize.

Step 3. Compare the reading on the glass thermometer with what the thermostat needle shows the temperature in the room to be.

Step 4. If the variation is more than a degree, check to see if your thermostat is dirty. Dirt can cause inaccuracies.

Step 5. Remove the face plate, usually held by a snap or friction catch, and blow away any dust with either your own breath or with a plastic squeeze bottle used as a bellows. Never use a vacuum cleaner because its suction force is too great.

Step 6. If your thermostat has

open contact points (those not sealed within a glass enclosure), rub a business card between them to clean these spots. Sandpaper is too abrasive and should never be used for cleaning contact points.

Step 7. If you have the type of thermostat with a mercury vial inside, make sure that the unit is level. If it is not, loosen the mounting screws and adjust the thermostat until it is level.

Some thermostats can be recalibrated. To find out whether yours can be, check the owner's manual or ask a dealer who han-

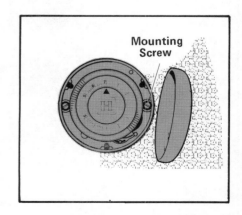

After taking off the face plate, remove the mounting screws in order to release the old thermostat from the wall.

dles your brand of thermostat. But if you have checked everything and determined that your thermostat is not calibrated properly, you should replace it. Be sure to buy one of the same voltage as your old one and one that is compatible with your heating system.

Low voltage thermostats can be installed safely without shutting off the unit's current, but it is nevertheless advisable to trip the circuit breaker switch or pull the fuse while making the changeover. Here is the basic procedure to follow when replacing a thermostat.

Step 1. Take the face plate off the old unit and look for the mounting screws. Remove the screws to release the thermostat from the wall.

Step 2. Remove the wires from the back of the old thermostat by turning the connection screws counterclockwise. Take care not to let the loose wires fall down between the walls.

Step 3. Clean the exposed wires by scraping them with a knife until the wire ends shine.

Step 4. Attach the wires to the new thermostat. The new unit must operate on the same voltage as the previous thermostat did.

Step 5. Push the excess wire back into the wall.

Step 6. Tape up the opening to prevent cold air inside the walls from affecting the thermostat.

Step 7. Install the mounting screws to secure the new thermostat to the wall. If the thermostat has a mercury tube—mercury tube thermostats must be level to be accurate—set the unit against a level during the installation.

Step 8. Snap the face plate back in place.

Step 9. Make sure that the new thermostat turns your heating/cooling system on and off when you alter the temperature setting.

Humidity

THE AMOUNT of humidity in your home can make a significant difference in your comfort. Too little humidity during cold weather can make you feel chilled, even though the temperature inside your house is high enough for comfort. Just the reverse is true during hot weather, when the amount of humidity bears an inverse relationship to how comfortable you feel. By regulating the humidity level in your home, therefore, you can achieve a pleasant level of interior heating or cooling without consuming a great deal of expensive energy.

"Relative humidity" refers to the percentage of moisture actually contained in the air as compared to the total amount it could hold. In other words, a relative humidity of 50 percent means that the air is holding half as much moisture as it is capable of holding at its present temperature. Since warm air can hold more moisture than cool air, a rise in temperature without a simultaneous increase in the moisture content means a decline in the relative humidity. If the temperature falls while the moisture content remains the same, the result is an increase in relative humidity.

A relative humidity of between 30 and 50 percent produces a pleasant level of comfort for most people. When the relative humidity falls much below 30 percent, you may—in addition to feeling chilled—find that your throat and nasal passages feel dry, encounter shocks of static electricity, and discover that furniture and other wooden objects in your house are drying out. On the other hand, when the relative humidity greatly exceeds 50 percent, you will probably see condensation forming on windows. You may even see this con-

densation run down the glass to the metal frame or wooden sill, causing rust in the former instance and rot in the latter.

An air conditioner acts as a dehumidifier, collecting condensation on its coils and then draining this water away. If you have an adequate air conditioning system, therefore, you need not worry about your home's humidity during the summer. Many homeowners, however, experience an excess of moisture during both summer and winter. A tightly sealed house will retain the moisture created by cooking, laundry, baths, and other moisture-producing procedures unless something is done to get rid of the excess. Exhaust fans or open windows do the job in many homes, but frequently the installation of a dehumidifier is required.

If you experience the opposite problem—too little moisture in the air, especially during the heating season—then you should install a humidifier. Some people dispute the claim that a humidifier can reduce heating costs, pointing to cost of electricity used to operate the humidifier. If by adjusting the humidity level, however, you are then able to turn the thermostat down several degrees, you can save a substantial amount of energy. Properly used, therefore, a humidifier can be an energy-saving investment.

In terms of size and type, there are small tabletop humidifiers, portable floor units, and units that are built into central heating systems. In terms of operating principles, there are evaporator humidifiers and atomizer humidifiers. The evaporator type has a water reservoir over which warm air passes, picking up water by simple evaporation. Some evaporator units feature revolving pads that place moist surfaces more directly into the path of the warm air coming out of the furnace, while others use a fan to draw warm air in and through an evaporator pad. The atomizer humidifier, in contrast, sprays a fine mist into the air. Both types are available for use as separate

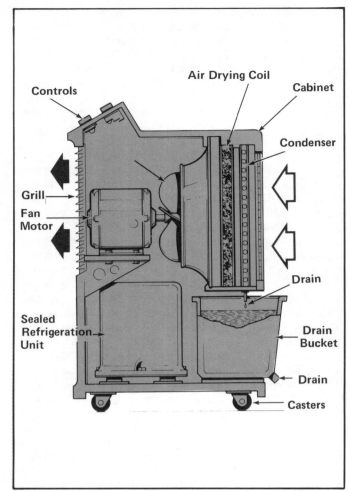

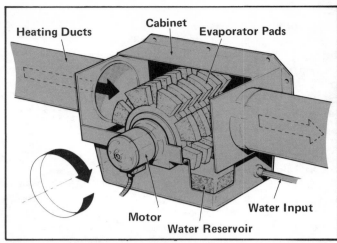

Some evaporator humidifiers feature revolving pads.

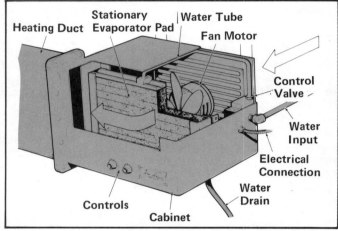

A portable dehumidifier can be positioned anywhere.

Others use a fan to draw warm air over an evaporator pad.

units or for installation in the central heating system. The best types are those that have a humidistat to control their operation according to the relative humidity inside the house.

Caring For Your Humidifier

MOST HUMIDIFIERS require routine maintenance in order to function at their best. If you follow these steps, you will optimize your humidifier's performance and keep it working for many years to come.

Step 1. Inspect the reservoir regularly since lime deposits from the water, plus airborne dust, can cause trouble. If you see any such deposits, clean the reservoir.

Step 2. Keep the media element (the pad, bristles, plate, or wheel) clean. Replace the elements when they can no longer be cleaned thoroughly.

Step 3. Inspect the water inlet valve for scale deposits.

Step 4. Check the float mechanism (the one that shows when the water level is low) frequently to be sure it is working properly.

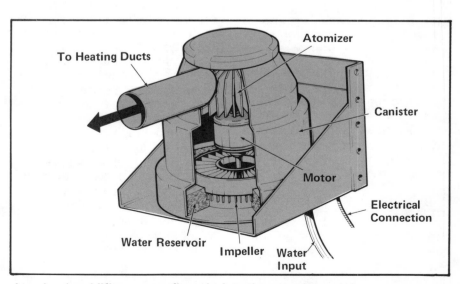

Atomizer humidifiers spray a fine mist into the warm furnace air.

Step 5. Always drain the reservoir at the end of heating season.

In addition to performing these routine steps, examine your owner's manual for care and maintenance procedures applicable to the particular type of humidifer you own.

Fireplaces

ONCE A necessity, the home fireplace is now a luxury. Not only are most fireplaces among the least efficient of all heat sources, but they also can waste the heat created by your conventional heating system.

Most of the heat created in a fireplace goes right up the chimney. In fact, many times the beautiful fire in the fireplace wastes far more of a home's heat than it produces. The reason for this phenomenon is that the fireplace depends on updrafts to keep the fire burning and to remove smoke. Therefore, the fireplace draws air from within the house —i.e., warm air from your furnace—and sends it up the chimney.

As much as 20 percent of the air in your house can be sucked up the chimney each hour. This represents a terrible waste and puts a big strain on your furnace as well as on your pocketbook. Even worse is the fact that after the fire is out, the updrafts can continue to remove warmed air from your house.

There are several devices on the market designed to reduce this energy loss and to improve the efficiency of the roaring fireplace fire. Some of these devices employ fans to draw room air through a chamber near the fire. The air is heated by the fire and then either blown back into the room or ducted into other areas. Some units accomplish the same

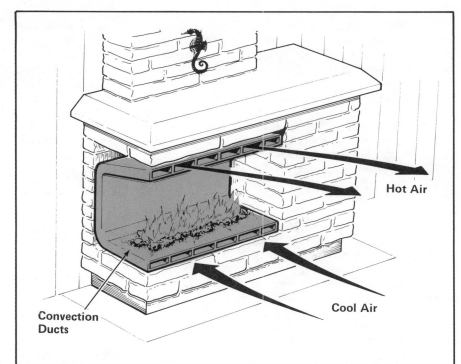

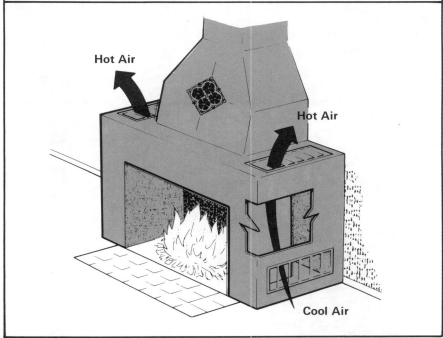

There are several devices available which can improve a fireplace's heating efficiency by taking in room air, heating it, and then sending the hot air back into the room. Some of these units have fans for better air circulation.

purpose without a fan. Air enters a hollow chamber, rises as the fire heats it, and then is directed back out into the room.

If you are planning to build a fireplace in a new home or add one to an existing home, examine the circulator devices which are designed to admit air through intake grills, carry it through hollow walls, and return it heated back into the room. Some of these units have fans to create even greater air flow.

At least as important as increasing fireplace efficiency while a fire is burning is decreasing the wasteful escape of warm air when the fire is out. A properly operating damper is, of course, essential to reducing this waste. If your fireplace lacks a damper, add one without delay, and then make certain that the damper remains closed whenever the fireplace is not being used.

One of the best ways to add a damper is to install a glass fireplace enclosure that features a built-in damper. Such an enclosure allows you to receive the fire's radiated heat and then to close both the glass doors and the damper, putting out the fire without losing any of your home's warmed air in the process. Some of these units also recirculate air while the fire is burning, and the glass doors, of course, serve an important safety function in preventing sparks from flying out into the room.

Fireplace maintenance is confined mostly to cleaning the chimney and to repairing masonry joints. Cracks in joints frequently cause bricks below to develop other cracks, a situation which can lead to the overheating of the surrounding wood structure and the threat of your house catching on fire. Follow standard procedures for repointing mortar joints, but if you must replace any bricks be sure to purchase fire bricks for this purpose.

Cleaning Your Chimney

THE HOME fireplace chimney usually requires little in the way of repairs, but you should clean it every so often. Although not one of your more pleasant jobs, cleaning the chimney will make your fireplace operate more efficiently. Here is how to make cleaning the chimney a do-it-yourself project.

Step 1. Open the damper.

Step 2. Seal off the fireplace opening from the room by attaching a heavy plastic sheet or a scrap piece of plywood with masking tape. Make sure that there are no cracks or leaks.

Step 3. Fill a burlap bag with straw, excelsior, or wadded paper, and then put in a brick or two for added weight.

Step 4. Fasten the bag securely to a rope.

Step 5. Climb up on the roof on a day you are certain that the roof is completely dry. Be sure to wear sneakers for good traction; some people tie a rope around themselves and the chimney.

Step 6. Lower the bag down one

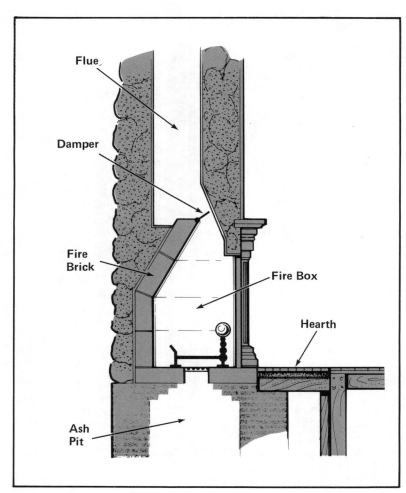

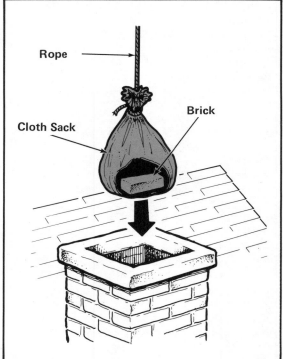

▲
Clean the chimney with a weighted bag filled with straw, excelsior, or wadded paper. Lower the bag into the chimney.

◀ *The home fireplace chimney generally requires little in the way of repairs, but you should clean it every so often.*

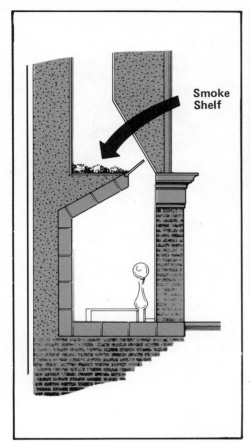

Smoke Shelf

▲
If your fireplace chimney has an outside access door, use it to remove soot loosened by the bag.

◀ *If you have a vacuum cleaner attachment that will reach up to the smoke shelf, use it to remove the accumulated soot and debris that collect there.*

corner of the chimney until it hits bottom.

Step 7. Raise the bag up and down several times.

Step 8. Now move the bag around the perimeter of the opening (move it about a foot each time) and repeat Step 7. Then get down off the roof.

Step 9. If your fireplace has an outside door, open it and remove the soot that you loosened.

Step 10. Wait for an hour or so while the dust settles before removing the plastic sheet or plywood covering from the fireplace opening.

Step 11. With the opening uncovered, take a large hand mirror and flashlight and hold them so that you can inspect the chimney. Look for any obstructions.

Step 12. Put on gloves, reach over the damper to the smoke shelf, and gently clean away the debris.

Step 13. Vacuum out the fireplace, and if you have an attach-

ment that will reach the smoke shelf, vacuum it as well.

Types Of Firewood

IN SELECTING what firewood to burn in your fireplace, you're generally limited by what is available in your section of the country; one specie may be in abundance, while others may not be available at all. If you have a choice, though, opt for the densest wood available. Generally the denser the wood, the better it is as fireplace fuel.

The following woods are listed according to their density. The difference between any two in the same category is insignificant, but the difference between a very high and a very low

wood represents a substantial disparity in value as fireplace fuel. You'll get more heat from woods higher up the list than from those in lower categories.

Very High

Live Oak
Hickory
Black Locust
Hackberry
Southern Yellow Pine
Persimmon
Apple
White Oak

High

Honey Locust
Black Birch
Yew
Blue Beech
Red Oak
Rock Elm
Sugar Maple
Ash
Black Walnut

Intermediate

Holly
Tamarack
Western Larch
Juniper
Red Maple
Cherry
Elm
Ponderosa Pine
Sycamore
Douglas Fir

Low

Magnolia
Red Cedar
Cypress
Chestnut
Black Spruce
Hemlock
Catalpa
Tulip Poplar
Red Fir

Very Low

White Spruce
Black Willow
Aspen
Butternut
Redwood
Sugar Pine
White Pine
Balsam Fir
Cottonwood
Basswood
Western Red Cedar

Cutting Your Own Firewood

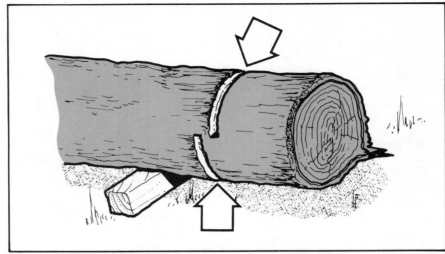

To prevent the saw from cutting into the dirt, cut the log from each side.

WITH THE ENERGY crunch, more people have started using their fireplaces as heat sources. Yet, firewood prices are so high that to achieve real savings many homeowners have taken to cutting their own firewood. Assuming that you can find trees to cut down, you can save yourself a good deal of money.

Of course, you must know what you are doing before you start felling trees. The following steps outline the basic procedure for turning trees into firewood.

Step 1. Determine which way the tree is leaning, and make certain that it won't fall on a house, power line, or anything else. Although you can control the di-

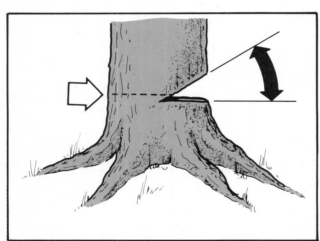

Saw a notch in the base of the tree on the side you wish to be the direction of fall. Make two cuts to create an opening of 45 degrees.

Prop the end of the log on another log or onto a sawbuck before cutting it into convenient lengths for home fireplace consumption.

The easiest way to split logs is with a wedge and a sledge hammer. Drive the wedge deep into the log.

rection it will fall to some extent by the cuts you make, you cannot compensate for the tree leaning in the wrong direction except by attaching a rope to the upper part of the trunk and having a helper guide the fall. Make sure that the helper stands far out of the way of the falling tree.

Step 2. Clear away brush from around the tree and in your path; you may need to get away quickly. Check for dead branches above that might fall on you, and if you spot any, remove them before you start cutting down the tree.

Step 3. Saw a notch in the base of the tree on the side you wish to be the direction of fall. You saw the notch by making a horizontal cut about a third of the way through the trunk and then making a diagonal cut downward at a 45-degree angle until you meet the end of the first cut.

Step 4. Now go to the other side of the tree and make the back cut a couple of inches higher than the first horizontal cut. Do not, however, cut all the way through to the notch you made.

Step 5. Remove the saw and put it out of your way.

Step 6. Drive a wedge into the back cut. Commercial wedges made of steel, aluminum, magnesium, or plastic are available or you can fashion your own from hardwood. Use a hammer to drive the wedge in, causing the tree to lean toward the under cut. When it leans far enough, its own weight will cause the tree to fall.

Step 7. Starting at the bottom of the tree and working toward the top, remove all the limbs. Cut off those that are sticking up, and then roll the tree (if possible) to reach branches on the down side. If the tree is on sloping ground, work from the high side.

Step 8. Cut the trunk into convenient log lengths, a procedure called bucking. To prevent the saw from cutting into the dirt, you need to have the tree trunk off the ground. You can prop the end on another log, or you can make a sawbuck—an "X"-shaped device that holds the

trunk. You should know ahead of time what size logs your fireplace can handle and have a branch or something measured to size so you can cut all the logs to the same length quickly and easily.

Step 9. You can split the logs either with a wedge and sledge hammer or with a maul or axe. The easiest way is with the wedge and sledge hammer. Stand the log on end and tap the wedge until it goes into the log; then drive it in with the sledge hammer. Once the log starts to split, remove the wedge, put the log on its side, and complete the split from the side.

Heat Saving Gadgets

THE FACT that many conventional heating systems are not as energy efficient as they should be has spawned dozens of gadgets designed to enhance efficiency and reduce skyrocketing fuel costs. If you recognize where the inefficiencies are in your system, you may be able to find products that help stop the waste of energy in your home.

Since the fireplace is probably the worst heat waster, check into some of the units on the market which are designed to improve fireplace performance. Convection grates, most of which have hollow pipes forming the log-holding portion of the grate, can help. These hollow pipes take in cool air at their floor-level openings. The air then rises as it is heated by the fire, passing through a "C"-shaped pipe until it comes out the top. Since the grate's top opening is aimed out toward the room, the hot air goes into the house instead of up the chimney. Some of these devices are also equipped with a small blower fan.

Fireplace hoods use a blower

to draw air in and through a heat exchanger over the fireplace grate. The heat exchanger is sealed to prevent combustion gases from being carried into the room.

Another source of wasted heat is the flue that carries combustion gases outside from the gas or oil furnace. There are several devices presently available that, when attached to round metal flues, remove the surface heat from the flue and release that heat into the room. Some of these flue attachments utilize a blower, while others let natural drafts move through to carry the heat away. Some attachments can even be installed with ducts to carry heat into other parts of the house.

A basement could possibly be made much warmer by this otherwise wasted heat, but before you invest in one of these devices, make sure that your flue is a good heat source to tap. Many flues are so well insulated that no heat is available for capture from the flue surface.

There are also gadgets available for diverting the flow of hot air from an electric clothes dryer. The devices filter the air and trap all the lint, retaining inside the house a substantial amount of warm moist air that would otherwise be ducted outside and wasted. These devices, however, are suitable only for electrically heated dryers since the waste air from a gas dryer also carries harmful combustion gases which must be vented outside.

Don't forget about the thermostat control devices that automatically turn the thermostat down at a preset time. Moreover, there are registers equipped with individual thermostats that permit homeowners to set a desired temperature in every room. The vents open and close automatically as the thermostat setting dictates.

Before you buy any of these devices—or any other of the new energy-saving products that are certain to appear in the months ahead—make sure that they pose no safety hazards and that they comply with all local building codes.

Heating

Materials and Supplies

Furnaces

Instead of a pilot light burning gas all the time, the Fedders Flexaire II Gas Furnace utilizes an electronic spark ignition. When the thermostat signals that it is time for the furnace to go on, an electrode produces a high voltage continuous spark that ignites the primary burner which in turn fires up the remaining burners. The Fedders fuel-saving electronic spark ignition system has a built in safety feature that shuts off the flow of gas in case of a power failure or a lack of ignition. (1)

The 397E gas forced-air furnaces from Bryant Air Conditioning are big energy savers—up to 20 percent during a heating season's operation—when equipped with the Vent Mizer and electronic spark ignition. The Vent Mizer is an automatic vent damper to keep heated air from escaping up the chimney between furnace firings. Other features include a coated heat exchanger for a long life, multi-speed blower, fan-limit control, printed circuit control center, slotted-port burners, and noise-insulated blower compartment.(2)

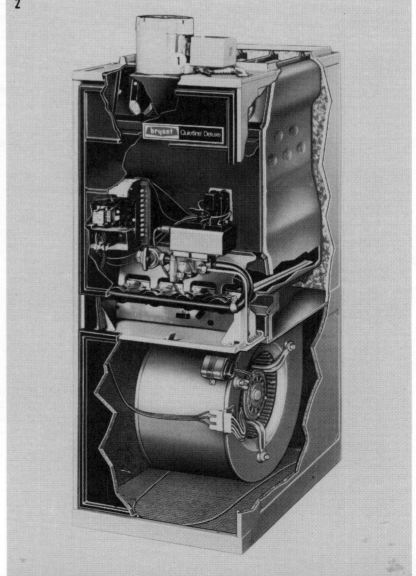

The Heat Recycler from Broan can reduce heating costs by as much as 15 to 20 percent. The unit eliminates thermal stratification in a room by drawing warm ceiling air in at the top and expelling it out the bottom where it mixes with cooler floor air. By breaking up the hot and cold layers of air, the Broan Heat Recycler permits a reduction of as much as five degrees in the thermostat setting while maintaining the same level of comfort. (3)

Amana Refrigeration produces a line of electric/gas cooling-heating units that feature a new Heat Transfer Module (HTM) heat exchanger. A product of aerospace research, the HTM was introduced in 1972 as a major breakthrough in heating technology. Small enough to hold in one's hands, this remarkable miniature heat exchanger is powerful enough to heat an entire home quickly and efficiently. It consists of a compact, stainless steel gas burner mounted in the core of a cylindrical heat exchanger. Fired by direct spark ignition rather than an energy-wasting pilot light, the stainless steel burner produces thousands of tiny flames that are clean burning and extremely hot. These flames create high temperature gasses which pass through and heat the finned surface of the heat exchanger. The fins, in turn, transfer the heat to an ethylene glycol solution circulating through steel tubes.

The combination wood and coal heater from Dornback Furnace and Foundry Company can either be attached to existing duct work or be installed as a separate heating system. Not just a space heater you can bring in and start using, however, it requires careful placement and must be connected to a masonry or class A all-fuel chimney. (5)

Cycle Pilot from White-Rodgers is designed for people who want to update their gas furnaces by switching from standing pilots to electronic ignition. White-Rodgers estimates that if all the gas furnaces with standing pilots were converted to electronic ignition, the total savings in natural gas consumption per year would amount to 295 billion cubic feet. Your local heating contractor can install the Cycle Pilot system for you. (4)

The Broan Millionair is a bathroom heater, exhaust fan, and light all in one. Any one or any combination of these functions can be activated due to the unit's two motor systems; one motor system operates a plastic centrifugal exhaust fan, while the other blows in heat from the 1500-watt

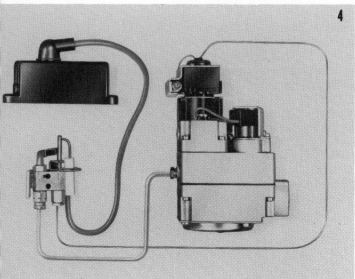

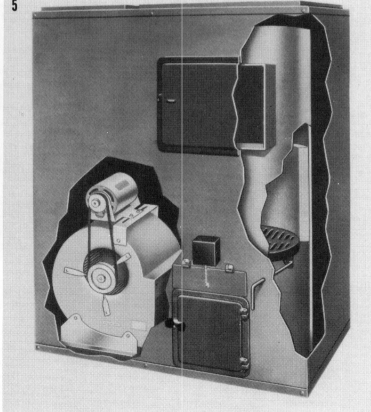

(1) Fedders Flexaire II Gas Furnace. (2) Bryant 397E forced-air furnace. (3) Broan Heat Recycler. (4) White-Rodgers Cycle Pilot system. (5) Dornback combination wood/coal heater.

(6) York Maximizer add-on heat pump. (7) General Electric Weathertron Heat Pump. (8) Janitrol Watt Saver heat pump. (9) Lennox HP10 split-system heat pump.

heater. A compact attractive unit, the Broan Millionair measures only 5¾ inches high and can fit into small ceiling spaces.

Heat Pumps

Carrier heat pumps, in various models and sizes, all feature the Weathermaker compressor which is an energy efficient de-vice with its own internal protection against excessive electrical current and heat. In addition, they offer the Chronotemp defrost unit which removes frost quickly when the outside temperatures go below 45 degrees and the AccuRater which meters refrigerant flow. Carrier heat pumps, moreover, come with aluminum coils for optimum heat exchange.

York's new add-on heat pump is called the Maximizer. Able to be added to most types of existing furnaces, the Maximizer can reduce electric heating costs by as much as 50 percent and gas or oil heating costs by as much as 20 percent. The Maximizer can handle chores as long as the outside temperature stays above 28 degrees. When the mercury dips below that point, however, the

Maximizer needs help from your furnace; at 10 degrees and below, your furnace does all the work. Like most heat pumps, therefore, the York Maximizer is at its energy-saving best in areas where winters are mild. (6)

The York line of Champion split-system residential heat pumps offer among the highest levels of heating and cooling efficiency in the industry. The units have energy efficiency ratios (EER) of 8.1 for the cooling cycle, and coefficient-of-performance (COP) of up to 3.1 (at 47 degrees) and 2.1 (at 17 degrees) for the heating cycle. Much of the credit for the high efficiency figures of the Champion heat pumps must go to York's unique solid-state logic module that controls the units' vital functions.

If you want a heat pump but are short on installation space, the Westinghouse Whispair Heat Pump may be just what you need. The tall slender aluminum cabinet is designed to be mounted on an outside wall with minimum protrusion. Available in three sizes, the Whispair is a quiet heat pump with a hermetically sealed compressor that is internally spring mounted for smooth operation.

Although many people think that the heat pump is a recent innovation in heating and cooling, General Electric has been making heat pumps since 1935. The Weathertron Heat Pump, a product of all those years of research and development, features GE's own Climatuff compressor which has a fine record for trouble-free operation. GE houses the components in the Weathertron in a galvanized steel cabinet with baked acrylic enamel. (7)

Lennox Industries recently introduced a new series of super-efficient heat pumps. The HP10 split-system units—currently available in 2.5, 3.0, and 4.0 ton capacities—are designed to produce cooling energy efficiency ratios of more than 8.0 and heating performance coefficients of nearly 3.0, placing the units among the best available for energy-efficient heating and cooling. An improved compressor and defrost cycle is largely responsible for the exceptional performance of the Lennox heat pumps. (9)

A heating-only heat pump is Janitrol's answer to the growing need for cost-effective electric heating systems. Rather than build a heat pump in which common components serve for both heating and cooling, Janitrol built one—called the Watt Saver—just for the heating function. The Watt Saver unit has a coefficient of performance (COP) of 3.2 at 47 degrees, which means that it will deliver 3.2 units of heat for every unit of electricity it consumes. The COP is 2.3 at 17 degrees. (8)

Mueller Climatrol Corporation makes a heat pump called Climator II which is powered by a rotary compressor for smooth, reliable operation. When used in combination with a forced-air furnace, the Climator II can be coordinated by Mueller's CliMizer Control Box, which protects the controls from dirt and moisture while keeping them accessible for servicing. The Climator II heat pump carries a limited one year warranty on all parts except the compressor which carries a five-year limited warranty; labor costs are not included.

Furnace Filters

Amer-Glas disposable filters fit all central heating and cooling systems. Available in hammock type, filter media rolls that can be cut to fit a particular unit, as well as individually pre-sized filters, Amer-Glas II filters (made by American Air Filter Corporation) have a media that allows you to tell at a glance when a change is needed.

While disposable air filters are the most popular, Phifer makes permanent Air-Weave filters that can be cleaned and reused.

These filters are all aluminum and can be cut with tin snips or scissors to fit your central heating/cooling unit. Like all permanent filters, they should be cleaned about once a month. Cleaning consists of a thorough vacuuming, followed by washing with soap and water and then spraying with a dust-attracting filter coater (such as Alfcoater) after they are dry. Permanent filters made of fiber-type material are also available, but the aluminum ones last longer and are easier to clean

The biggest enemy of central heating and air conditioning equipment is dirt. A clogged filter robs the unit of up to five percent of its efficiency. Honeywell offers a convenience device that indicates when the filter is clogged and needs changing or cleaning. It can be installed in the furnace's blower compartment by professional heating personnel or by a handy do-it-yourselfer.(10)

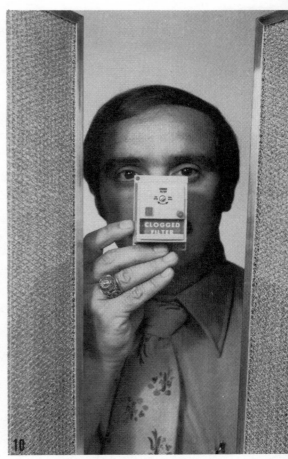

(10) Honeywell clogged-filter indicator.

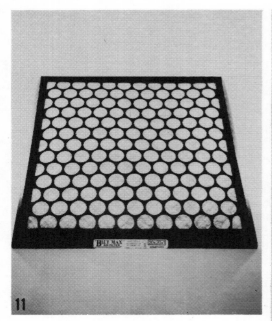

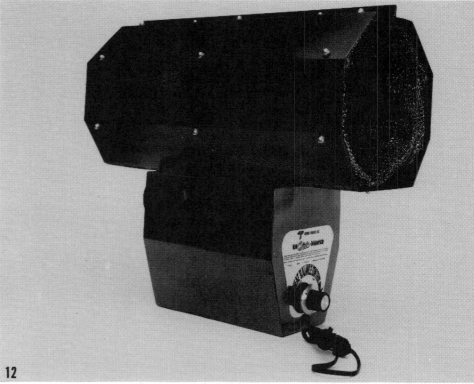

(11) U.S. Gypsum Blue Max air filter.
(12) Recyclo-Heater. (13) Isothermics
Air-O-Space flue device. (14) Vent
Mizer vent damper.

Blue Max air filters from United States Gypsum Company come in a convenient six-pack. One package should get you through a complete heating season. Blue Max glass fiber filters come in the most popular sizes.(11)

Wood Burning Stoves

Called the Cannonball, the pot-belly stove from the Washington Stove Works is made of solid cast iron and comes with a lifter and a cooking lid. Although the Cannonball gets very hot, it's strictly a room heater unless an auxiliary fan is installed to circulate the heat it produces.

The handsome parlor stove replica from Hearth Craft comes in a choice of three exterior trim designs and has dual damper controls to regulate the air flow through the wood fire. Hearth Craft makes several other cast-iron stoves plus a line of free-standing fireplaces for do-it-yourself installation.

An authentic replica of the old west range, the Sheeprancher—a creation of U.S. Stove—features a cast-iron top and cooking surface and comes with a lid lifter. It can burn wood or soft coal.

Furnace Flue Devices

The flue stack from the furnace carries a good deal of heat away in addition to harmful combustion gases. The Recyclo-Heater is designed to attach quickly around the flue pipe and then to absorb a significant amount of this escaping heat. As the temperature rises, the unit's thermostat activates a blower which draws air through the Recyclo-Heater. The air is heated and then either sent out into the room or ducted to other areas. The Recyclo-Heater can be installed easily without cutting into the flue pipe, but the cost of the unit is such that it would take years of heat savings to recover the investment.(12)

The Air-O-Space unit from Isothermics captures heat that

would otherwise escape up the flue. As the heat rises, it warms special pipes in the Air-O-Space. These pipes then warm the room air. The Isothermics Air-O-Space fits 6-inch, 7-inch, or 8-inch flue pipes.(13)

The Handyman's Heat Recovery System from Dutch's Enterprises, Inc., fits any size flue pipe and can be installed easily in minutes. It consists of strips of heat-absorbing aluminum fins which transfer heat to the air by convection. Installation involves attaching the first strip to the flue with a self-tapping screw and then just wrapping the rest around the flue.

A Stack Pack unit from Flair Manufacturing Corporation can save fuel when installed on an oil-fired furnace, boiler, or water heater. As soon as the burner unit shuts off, a signal goes to the Stack Pack which closes the damper to stop standby heat loss up the chimney. A time delay eliminates combustion odors. Stack Packs are available in standard sizes for 6-, 7-, 8-, and 9-inch flues, but these devices

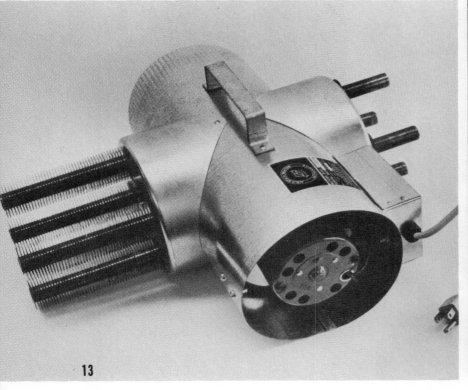

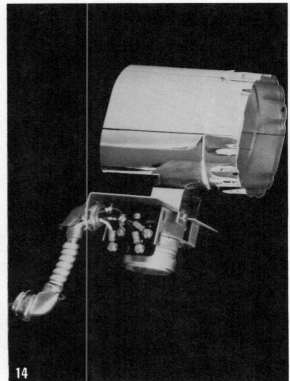

are to be used only with oil-fired burners.

A unique automatic vent damper, called the Vent Mizer, has been introduced for use in Bryant's 397E series of gas-fired furnaces. The energy-saving device mounts on the furnace's flue gas discharge and automatically prevents heat from escaping up the chimney when the furnace is not operating. It does this by closing the flue when the burner goes off to lock in both the heated indoor air and the hot air trapped in the exchangers. (14)

When the furnace is going, the Heatkeeper lets combustion gases escape through the flue pipe. But when the furnace shuts off, the Heatkeeper closes the flue and keeps heat from rising up and out the chimney. The Heatkeeper automatic flue damper is wired into the furnace so that it will not close when the furnace is on. Manufactured by Maid-O'-Mist, well-known makers of humidifiers, the Heatkeeper can be installed easily in the existing flue pipe.

Magic Heat is a unit that fits into metal flue pipes of a heating unit to capture heat that normally goes up the chimney. As heat goes through the unit, passing across heat-collecting tubes inside the heat exchanger, it causes a thermostatically controlled blower to come on. The blower brings room air into contact with the hot tubes, and then blows it back out into the room as heated air. Magic Heat, from Calcinator Corporation, has a soot cleaner which facilitates the important task of keeping the tubes free of deposits.

Calcinator Corporation has taken the idea of recapturing heat lost up the flue one step farther. A Granny's Oven unit, which fits into the flue pipe by replacing a regular 18-inch section, captures a good deal of heat while allowing combustion gases to escape. In fact, this unit captures so much heat that it becomes a handy oven for baking bread, biscuits, and cookies; there is even a thermometer in the door to tell you the inside temperature. Although made only for 6 inch flues, Granny's

Oven can be equipped with optional adapters to let you fit it to other flue sizes. The recaptured heat of course, also helps warm the house.

Thermostats

The Thermotimer from the Thermotrol Corporation is a battery-operated thermostat control that fits over most conventional round thermostats. You simply set the Thermotimer for the low and high temperatures you want and the times you want the changes made.

In a centrally heated and cooled home, there are always rooms that are either too warm or too cool. This familiar problem can be eliminated with Trol-A-Temp registers or dampers. These motorized devices have separate automatic thermostats to send heated or cooled air to the various rooms as needed. The thermostat in each room can either open the damper or close it, and it can turn the entire heat-

ing/cooling system on or off, naturally, a Trol-A-Temp can shut off the central system only when the thermostats in the other rooms do not signal the need for additional heating or cooling. Trol-A-Temp registers and dampers come in different styles to fit just about any type of central heating/air conditioning system.

Thermostats do wear out and must be replaced from time to time. When you face this situation, Jade offers a wide variety of replacement models to choose from—vertical, horizontal, and even one round unit—for natural gas, electric, propane gas, or oil systems. Jade thermostats are accurate, decorative, and easy to read, and they feature positive off settings.

Ammark Corporation markets a line of Fuel/Saver room and zone thermostats, all featuring temperature setback capabilities. The Ammark 7-day automatic clock thermostat combines a 24-volt room thermostat and a 24-volt 7-day electric clock, which can be programmed for various daily temperature setback periods. Degrees of temperature setback are adjustable within a range of zero to 15 degrees Fahrenheit.(15)

The Deluxe Regutemp II combines a good thermostat with a solid-state quartz crystal automatic regulator to allow those with central heating and air conditioning systems to designate two timer settings; for daytime and nighttime. The Deluxe Regutemp II, a product of General Time Corporation, is designed to replace most low voltage thermostats. Another version is available that works only with the heating system. Installation is an easy do-it-yourself project.(16)

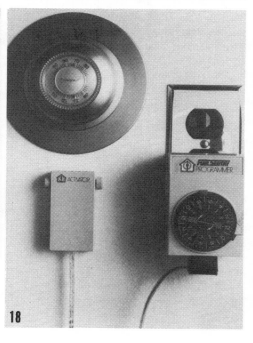

(15) Ammark Fuel/Saver. (16)General Time Deluxe Regutemp II. (17) Fuel-Miser. (18) Fuel Sentry.

The Fuel-Miser is a do-it-yourself control that converts any wall thermostat to an automatic clock thermostat in minutes. The UL-listed device saves fuel by allowing users to program reduced levels of heating or air conditioning at bedtime or when they are away from home. It automatically restores comfortable temperatures before they awake or return. No wiring to the thermostat or to the heating or cooling equipment is needed. One component plugs into an electric outlet, and the other screws to the wall beneath the existing thermostat. The clock is then programmed for the time that heating or cooling is to be reduced and for the time that comfortable temperatures are to be restored. The unit then repeats its function daily without resetting. On those occasions when the user may not want heating or cooling reduced at the programmed time, a switch on the Fuel-Miser turns off the thermostat control but does not affect the clock. There's also a built-in transformer that allows the device to operate at a safe, economical low (24V) voltage.(17)

The Fuel Sentry is a device you can add to your thermostat to turn the heat up shortly before you get out of bed in the morning and reduce the heat automatically at night. During the summer, Fuel Sentry can cut cooling costs by turning off the central air conditioning when no one is at home, and then turning it on in time to get the house cooled off before it is occupied once again. The Fuel Sentry, a product of the Fuel Sentry Corporation, is easy to install.(18)

Fireplace Heating

The Thermograte (from Thermograte Enterprises, Inc.) is one of those fireplace grates that circulates cold air in and around the fire through hollow tubes and then returns the heated air into the room. The tubes are made of heavy 14-gauge tubular steel welded to heavy steel cross bars that form the legs and upper support. Thermograte comes in a number of different sizes.

The Super Energy Grate, a product of TechnoSci Inc., employs a blower to increase air circulation. In addition, each Super Energy Grate comes with a firebox and heat reflector; a battery supply unit for emergency power is available as an option. Another

option is an automatic timer to shut off the blower at a desired time. Both the blower and the power cord are adequately protected from excessive heat, and none of the equipment requires any installation other than placing it in the fireplace and plugging in the blower cord. TSI also markets a Mini Energy Grate (designed to fit many free-standing fireplaces), Franklin Stoves and other small heating units and fireplace devices.

The Free Heat Machine is of course, not really free. In addition to the purchase price, you must still pay for the electricity its blower fans consume. On the other hand, this device from Aquappliances, Inc., does save much of the heat that normally escapes up the fireplace chimney. The Free Heat Machine is larger than the fireplace opening and extends into the room by about eight inches. Twin blowers are at either side, and the heat exchanger is bigger than most of the other wood grate devices designed to utilize fireplace energy.

A spark screen is also provided, and filler panels are available to close gaps above the unit.

Hot Air Andirons by Lauderdale-Hamilton is a unit that utilizes a small blower to bring cold air from the room, convey it through the hollow andiron, and then expel the heated air out into the room. Constructed so that there are no holes in the fire area that could pick up smoke or combustion gases, the Hot Air Andirons also has no welds in areas where heat might undo them. Although the unit is not as effective as the multi-tube wood grates designed to circulate air around a fireplace fire, it is considerably less expensive. (19)

Convect-O-Heater utilizes a small silent blower to force air through hollow tubes that form a fireplace grate. This forced convection moves the air farther out from the fireplace than similar units that rely upon natural convection. No installation is required; the unit merely replaces the standard grate. Available in

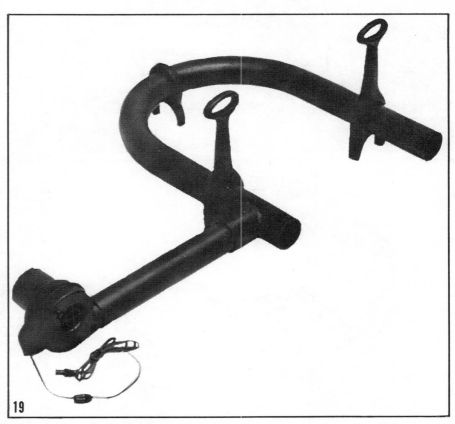

(19) Lauderdale-Hamilton Hot Air Andirons.

several models, including some with variable blower speeds, the Convect-O-Heater (from General Products Corporation) also comes with a glass doored screen unit that has openings at the top to allow the heated air to come out into the room. (21)

Duo-Therm's Hearth Heater consists of an attractive hood that mounts at the top of the fireplace opening with self-tapping screws or clamps. Inside the unit are two heat exchangers sealed so that flue gases can't be blown into the room—blower fans, and an automatic thermostat. The Hearth Heater is available in two sizes to fit most masonry fireplaces. (20)

A free-standing prefabricated fireplace called the Provider has a built-in fan to circulate heat. It can also be equipped with an optional outside air intake kit, called Energy Mizer, that prevents the loss of warm room air through updrafts. The Provider, from Preway Inc., comes in a choice of black, red, or olive green. (22)

With a little work, a metal drum, and a Warm-Ever kit from Locke Stove Company, you can fashion your own wood-burning barrel stove. The kit includes the sturdy cast-iron legs, the cast-iron front feed door with draft slide, and the flue outlet piece. It burns firewood up to 24 inches in

length. If you aren't handy, but like the idea, Locke Stove Company markets the complete unit assembled and lined with firebrick. (25)

A new energy-efficient, built-in fireplace has been introduced by The Majestic Company. Called the Energy Saving Fireplace, it features a stainless steel heat exchanger and stainless side liners for substantially increased radiant heating benefits. Advanced engineering concepts are used in the heat exchanger unit in order to expose a large amount of compactly arranged surface area directly to the fire. In addition, the Majestic fireplace has built-in room air outlets and intakes for

20

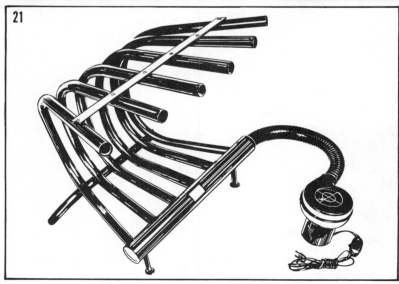

21

22

optimum warm air distribution, and can operate with or without optional blower fans. Its innovative design allows the Energy Saving Fireplace to transfer convected heat effectively into the room. (23)

The Protecto-Pane glass fireplace enclosures from Bennett-Ireland come in many sizes and styles to match almost any decor. More importantly, these screens are designed to prevent fireplace updrafts from removing large quantities of warmed air from your home. The Protecto-Pane includes a damper that can be closed when you wish to retire and lets the fire burn out without the loss of warm air from the house. The tempered glass doors prevent an ember from escaping and posing a fire hazard.

Firewood Tools

The advent of chain saws has made cutting firewood an easy task for the average homeowner. The lightweight McCulloch Mini-Mac saws are light enough for anyone to handle, but they can handle tasks around the house. For example, the Mini-Macs with a 12-inch bar can cut up trees 24-inches thick and do it quickly. If you use a good deal of firewood, an inexpensive chain saw like the Mini-Mac will pay for itself in a single season.

The J.C. Penney 14-inch electric chain saw, which weighs less than 10 pounds, is good for felling small trees and converting them to firewood. It has a centrifugal switch to prevent overloading the motor and a security switch to prevent accidental starts. The housing is double insulated.

Beaird-Poulan, a division of Emerson Electric Company, has introduced a lightweight chain saw designed and priced for the home user. The Micro Super 25 Deluxe weighs less than 10 pounds, but a 14-inch sprocket nose bar makes it possible for the saw to fell a tree 28-inches in diameter. It also has a cush-

(20) Duo-Therm Hearth Heater. (21) General Products Convect-O-Heater. (22) Preway Provider Fireplace. (23) Majestic Energy Saving Fireplace. (24) Homelite Safe-T-Tip. (25) Locke Stove Warm-Ever wood-burning stove. (26) Beaird-Poulan Micro Super 25 Deluxe.

ion-grip handlebar, a low-profile chain, built-in safety link, and a quiet spark-arresting muffler. The unit's reed valve induction system results in a cooler running carburetor and a high voltage, moisture-and-dirt-proof ignition system.(26)

A hardened steel protective device called Safe-T-Tip has been developed by the Homelite Division of Textron Inc., to eliminate the danger of kickback associated with chain saws. Safe-T-Tip, which weighs just an ounce, fits over the nose section of the guide bar to cover the sensitive area of the bar and chain where kick-back reaction is generated. The device is attached with a mounting screw of high tensile strength that can be installed or removed in a few seconds. Offered free on Homelite consumer chain saws, Safe-T-Tip is available as an add-on accessory for most other makes. (24)

The Pitch-N-Gauge is a flexible plastic card with holes, slots, and cutouts to check bar groove file sizes, tooth length, and drive links, set filing angles, and the depth of cutters. A creation of Granberg Industries, Inc., Pitch-N-Gauge measures less than six inches long and is available where chain saw supplies are sold.

Since split logs burn better, you still have some work to do after you've finished cutting your firewood. The double-faced sledge, from the Warren Group, is made of high-carbon heat-treated steel, has machined and polished faces, and is available in 6-, 8-, 10-, 12-, 16-, and 20-pound sizes. Warren also makes a line of wedges, including the popular 6-pound Oregon Splitting Wedge.(27)

The Blue Ox Log Splitter consists of a splitting wedge on a bar suspended in a frame that holds both log and wedge in place for your sledge hammer blows. The device can handle logs up to 12 inches in diameter and does the splitting quickly and safely. The Blue Ox is made by Dasco Prod- ucts, makers of conventional splitting wedges in all the popular sizes.(28)

You can split firewood fast and easy with the WoodSplitter and Wedge from Woodings-Verona. The WoodSplitter can handle small to medium logs and softwoods by itself, but splitting larger logs and hardwoods requires the square head wedge. If you've never cut firewood before, Woodings-Verona markets a WoodSplitter kit complete with all the tools and information you need to become adept at firewood splitting.

Fireplace Cleaners

If you're planning to build a fireplace, be sure to include an ash pit. Heatilator Fireplace Division of Vega Industries makes a good ash pit that consists of an ash dump unit and a cleanout

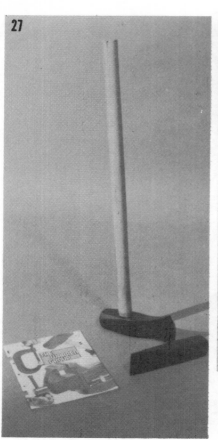

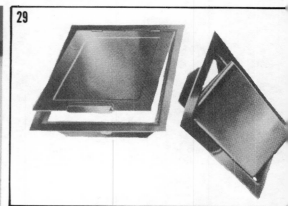

(27) Warren Group sledge. (28) Dasco Blue Ox Log Splitter. (29) Heatilator ash pit. (30) Coughlan Chimney Sweep Fireplace Powder.

door. The ash dump, installed in the hearth, allows ash disposal into the cleanout pit. The cleanout door makes ash removal possible from outside or from the basement. (29)

To get rid of soot in the chimney, sprinkle Chimney Sweep Fireplace Powder (made by Coughlan Products) on the burning logs in your fireplace. As it burns, the powder emits a chemical soot remover that goes up the chimney. As a bonus, the powder burns in a brilliant rainbow of colors while cleaning your chimney. (30)

Humidifiers

The tiny Thermo-Mist humidifier attaches to the central heat ducts and atomizes moisture into the air. The standard nozzle provides about nine gallons of moisture per day, and optional nozzles are available to increase the output; a control knob regulates the output. Household water pressure is used to inject the water, and a thermal sensor allows the unit to operate only when the furnace is blowing out enough hot air to evaporate the spray. The solid-state Thermo-Mist has no motor, no floats, and no reservoir to clean. It can be installed by any reasonably competent do-it-yourselfer. (31)

A console humidifier can provide an ideal way to add moisture to your home if you live in an apartment or if your method of heating doesn't lend itself to installing a built-in unit. Installation of a console humidifier requires only plugging it into a wall outlet. The Comfort-Aire humidifier, which comes in a Mediterranean-styled cabinet made of wood-grained vinyl on steel, has an automatic humidistat, a water level indicator with an indicator light to show when the tank is empty, and an automatic shutoff. The fan has a variable speed control and can dispense up to 12 gallons per day.

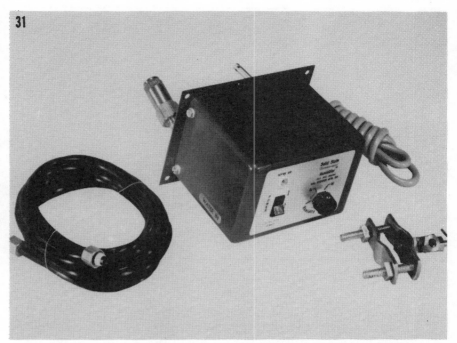

(31) Thermo-Mist humidifier. (32) Good-Bye-Dry humidifier.

A whole-house humidifier designed and engineered expressly for do-it-yourself installation is being marketed by Comfort Enterprises. The unit, called Good-Bye-Dry, mounts under the warm air duct of any forced air heating system. With a rated capacity of 14 gallons per day, Good-Bye-Dry comes pre-plumbed and pre-wired, so that installation entails merely cutting a rectangular opening in the heating duct, mounting the unit, inserting the self-tapping saddle valve in the nearest cold water pipe and plugging the 24-volt transformer into any household socket. All the necessary hardware is included. (32)

Air Conditioning

AIR CONDITIONING is no longer considered—as it once was—a luxury item. Instead, it is an essential part of summer living for millions of Americans. Many new homes come equipped with central air conditioning, and thousands of older homes are being converted to central air every year. In addition, window units (more correctly called room air conditioners) are also very popular. Easily installed—many require no special tools or extra wiring—these room units come in a variety of cooling capacities, from portable models to units capable of cooling several rooms.

The type of air conditioning system that's best for you depends on your needs, the type of home you have (i.e., does it have ductwork for central air?), and how much money you wish to spend. To make the most energy efficient choice, however, you need to have additional facts about the various air conditioning systems currently available.

Central Units

IF YOU decide to have central air conditioning installed in your home, it is essential that the system you select be of the correct size. A unit that is too small must run all the time to keep pace—an expensive and energy-wasting situation that may never keep your home comfortably cool. A unit that is too big, on the other hand, is also wasteful. An oversized unit costs more to operate due to excessive cycling. It cools the house quickly and then shuts off, only to come back on a short time later. The extra energy consumed each time the unit comes on is costly, and the excessive cycling is very hard on the unit's operating mechanisms. In addition, an oversized unit can cool so fast that it does not properly dehumidify your home, leaving you cool but clammy.

The best way to learn what size unit to install is to ask an air conditioning engineer. He will survey your home and prescribe what is needed. What factors does the engineer take into consideration? The square footage of living space is fundamental, although there are many other things that must be taken into account. The number of windows, their size, which way they face, and the type of glass in them; the way the house is laid out (air flow and circulation); the type of construction; the amount of insulation; the number and nature of the appliances, the amount of their use, and their wattage; and, finally, trees, shrubs, terrain, and other outside influences—all of these factors have a potential bearing on the size of the central air conditioning unit required.

It's a good idea to seek three different estimates on the size needed from three reputable air conditioning engineers. All of the estimates should be quite similar, and you can verify the engineers' findings by checking the tables published by air conditioner manufacturers to help prospective customers estimate their own needs.

Unfortunately, however, there is considerable confusion about what determines the size of an air conditioner. Many people refer to a unit in terms of "tons." This tonnage doesn't refer to the unit's weight, of course; rather, it refers to the tons of ice that would have to be melted to produce equivalent cooling.

The more meaningful way to specify air conditioner size is in BTU's (British Thermal Units). A BTU rating refers to how many of these units of heat a given unit can remove in a one hour period.

Since a one ton air conditioner can remove 12,000 BTU's per hour, conversion from one type of specification to the other is quite simple.

A horsepower rating cannot be converted either to tonnage or BTU since units with the same horsepower can differ significantly in cooling capacity. Consequently, horsepower is a meaningless air conditioner specification.

After size, your next consideration in selecting an air conditioner is efficiency. Even though two units have the same BTU rating, one may well have a greater energy efficiency ratio (EER). Once you find the BTU rating, look at the plate that shows the unit's wattage. By dividing the wattage into the BTU rating you get the EER.

$$\frac{BTU\ Rating}{Wattage} = EER$$

For example, a 12,000 BTU air conditioner requiring 1600 watts has an EER of 7.5 (12,000 ÷ 1600 = 7.5). This means that the unit will consume 1600 watts per hour in providing 12,000 BTU's of cooling (one ton of refrigeration). A unit of the same size that consumes just 1200 watts would have an EER of 10 and, therefore, would be a more efficient unit. The higher the EER the more efficient and less energy-consuming the unit. Thus, while a unit with a high EER may cost more, it will usually pay the difference back in energy savings.

Installing A Central Air Conditioner

AFTER YOU determine the size unit you need and purchase a model at the price and EER you desire, you must make certain that the central air conditioner is properly installed. Here are some points to consider when planning the installation.

1. The outside condenser unit should be located where it will receive a minimum of direct sunlight. If there is no good location, you should provide shade with a fence, trees, or shrubs.

2. The outside fan should not blow against a wall. Otherwise, the hot air that had been discharged would blow right back into the air intake.

3. If the air intake must face a wall, make sure that at least twelve inches separate the intake from the wall.

4. Plan the location of your air conditioner unit so that it will not interfere with your patio or other outdoor living areas. Even the quiet units are still disturbingly noisy. In addition, they all offer a certain amount of vibration which could cause problems if the unit is improperly mounted adjacent to your home.

5. If you already have a central heating unit, check the adequacy of the ductwork for air conditioning. Cool air does not travel as rapidly as hot air through small ducts.

6. All ducts should be properly insulated.

7. Provide a separate electrical circuit for the central air unit.

8. If you are adding air conditioning to an existing central heating system, you generally have no choice as to where to put the blower. If you do have a choice, however, place it where it will be able to move the air throughout the house.

9. Position the blower unit so that it will be easily accessible for maintenance.

10. Position the refrigerant lines so that they are protected from accidental damage.

11. Be sure your house is properly protected from condensate.

12. Make certain that the system will have ample return-air vent space.

13. Position the thermostat so that it provides an accurate reading of your home's temperature.

14. Be sure that the installation complies with all local codes.

The good installer will take care of all of these matters and perform any special installation procedures needed to provide you with greater efficiency and comfort.

Maintaining And Repairing Central Air Units

MOST OF THE maintenance that a central air conditioning unit requires to operate at its best concerns cleanliness. Like a central

Common Sense Tips On Getting The Most From Your Air Conditioning

1. Aim the vents of room air conditioners upward for better air circulation. Cold air naturally settles downward.

2. If you have room units, close all heating system vents so that the cool air isn't wasted going inside the walls and down through the ducts to the basement.

3. Lighting fixtures throw off more heat than you'd imagine; don't have them on unless absolutely necessary.

4. Use shades, blinds, and drapes to keep direct sunlight from entering your home.

5. Keep heat-producing devices far from where they might influence an air conditioner thermostat.

6. Make certain that the outside portion of the air conditioning system, whether a room or central unit, is neither in direct sunlight nor blocked from free air flow.

7. Keep furniture and drapes and other obstacles out of the way of air conditioning vents.

8. Turn on exhaust fans to remove moisture from the kitchen and bath, but turn them off as soon as the warm moist air is out.

9. If you're away from home every day, install a timer control to keep the air conditioner off until shortly before you get home in the evening.

10. If you're going to be away for several days, turn off your central air conditioning system.

11. Keep the system properly maintained and in good working order.

heating system, the air conditioning system must be kept clean, and that means changing filters frequently.

In addition to the filters, the evaporator inside and the compressor/condensor unit outside should be kept as clean as possible. Frequently, the evaporator is not accessible, but if it is, you should clean it on an annual basis.

The evaporator is located just forward of the furnace in a small sheet metal box called a plenum. If the plenum has foil-wrapped insulation at its front, you can proceed to clean the evaporator; but if the plenum is a sealed sheet metal box, do not attempt to open it.

If you find that you can reach and clean the evaporator, here are the steps to follow.

Step 1. Cut the electric power to the unit by tripping the circuit breaker switch. Make sure that the current is off before you start working.

Step 2. Remove the foil-wrapped insulation at the front of the plenum. It is probably taped in place. Remove it carefully because you will have to replace it later.

Step 3. Once the insulation is off, you will see the access plate which is held on by several screws. Remove the screws and the plate will come off.

Step 4. In some cases, you will now be able to reach all the way back to clean the entire area. If not, slide the evaporator out a little. You can slide it even though the evaporator has rigid pipes connected to it, but be sure not to bend the pipes.

Step 5. Use a suede brush to clean the entire underside of the evaporator unit. A large hand mirror can be very helpful in letting you see what you are doing.

Step 6. Clean the tray that is located below the evaporator unit. This tray carries away condensation from the evaporator, and if it gets dirty it can clog up. Next, pour some household bleach into the hole in the tray to get rid of any fungus that may be forming.

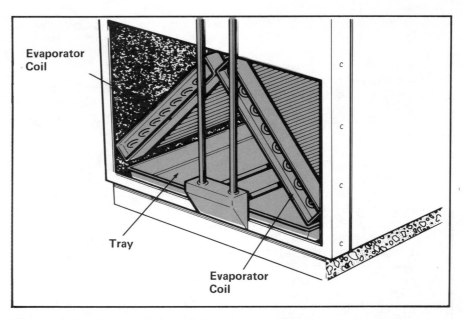

The tray below the evaporator coil is there to catch and carry away condensation. It must be kept clean and free of fungus growth or else it can clog up. After cleaning, pour in bleach to kill any fungus spores.

Step 7. Put the evaporator unit back in place, reinstall the plate, and tape the insulation back over it. Check for any air leaks and cover with duct tape.

The outside cleaning must be done at the condensing unit. The unit has a fan that moves air across the condenser coils. Since you must clean the coils on the intake side, check while the unit is running to ascertain which direction the air moves across the coils. After you know this, then you can start cleaning.

Step 1. Cut all the electric power to the unit by tripping the circuit breaker switch. Make sure that the current is off before you start.

Step 2. Cut down any grass, weeds, or vines that have grown

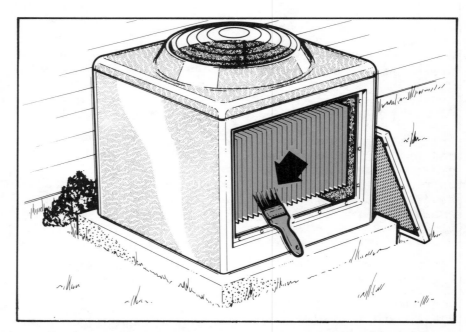

Use a brush to clean away accumulated dirt from the fins around the condensing unit. Work carefully; the fins can be damaged easily.

98

up around the condensing unit. They could be obstructing the flow of air.

Step 3. Use a brush to clean away all the dirt that has collected on the fins. Brushing works better than spraying with a garden hose because water can turn part of the dirt into mud and compact it between the fins. But be careful. The fins are made of light-gauge aluminum and are easily damaged. In many cases you will have to remove a protective grille in order to reach the fins.

Your air conditioning system is charged with a refrigerant called Freon. If the system lacks the proper amount of Freon, little or no cooling will take place. If you suspect a Freon problem, call in a professional to recharge the system. Never attempt to charge your own Freon lines unless you have the proper gauges, equipment, and experience to perform this procedure properly. Freon can be dangerous if you don't know how to handle it.

Room Air Conditioners

ROOM UNITS come in many sizes and styles, from single-room models to powerful multi-room cooling machines. Many of the smaller room air conditioners can be plugged into a convenient 120-volt wall outlet, while most of the larger units require a special 240-volt circuit of their own.

As with central air units, room air conditioners should be selected on the basis of the space to be cooled. A machine that is too small is a waste of money, while a powerful unit designed to take care of several rooms is a waste of energy dollars if the cool air it produces can't circulate through all the rooms. In the latter instance, a homeowner would be wise either to position a

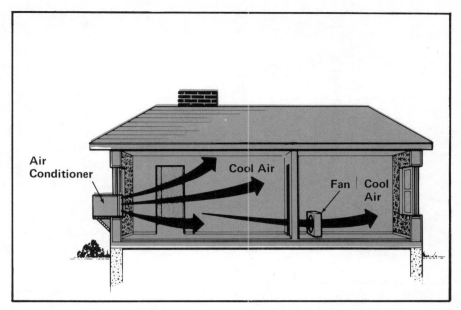

One room air conditioner can frequently be made to cool several rooms by careful positioning of a separate fan to help circulate the cooled air. Of course, the cooling unit itself must be one of substantial capacity.

separate fan to help move the air into the other rooms or to install two smaller units instead of the one big one.

Air circulation is not the only factor that affects cooling capacity. The number of windows in a room, which way they face, the room's insulation and weatherstripping, and even any heat-producing appliances in the room must be considered before an intelligent determination of room air conditioner size can be made.

After you decide on the size unit you need, seek out the one offering the best EER. A unit with a high EER frequently costs more than less efficient units, but you will recover the extra expense in energy savings.

Installing A Room Air Conditioner

UNLIKE A central air system, most room units are designed to be installed by the consumer. Generally, people place room air conditioners in windows, but the units can be installed through the wall as well.

Whether you install your room air conditioner in a window or through a wall, you must first consider some important facts regarding its location. The portion of the unit that sticks outside contains the condenser, and the

condenser needs to have good air circulation moving around it; otherwise, the heat drawn from inside won't be dissipated. The cooler the air around the condenser, the more efficient your unit will operate.

Accordingly, the unit should not be placed where it will receive a great deal of direct sunlight. If you have no choice as to location, consider mounting an awning above the unit to shade it from the sun.

Most room air conditioners come with instructions for window mounting. Some even come with an installation kit that includes the metal mounting frame, the cradle, side wing panels, and gaskets for sealing. While several models require special mounting methods, you can mount most in double-hung windows by following these steps.

Step 1. Assemble the mounting hardware.

Step 2. Attach the frame cradle assembly to the sill. Since this procedure can vary according to the type of unit and the type of frame, be sure to follow the instructions provided in the mounting kit.

Step 3. Adjust the frame so that the unit will slant downward toward the outside. This position-

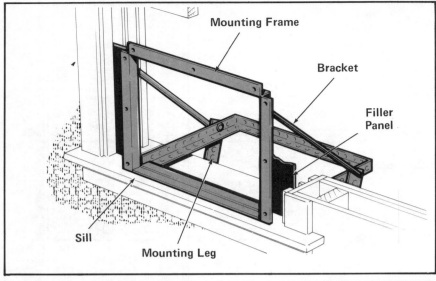

Mounting Frame

Bracket

Filler Panel

Sill

Mounting Leg

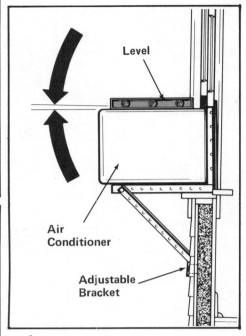

◀ *Attach the mounting frame assembly to the window sill according to the directions supplied with the unit.*

Level

Air Conditioner

Adjustable Bracket

▲

Adjust the frame so that the air conditioner will slant downward toward the outside.

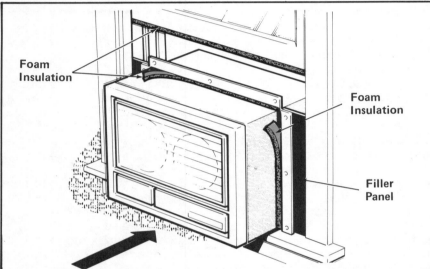

Foam Insulation

Foam Insulation

Filler Panel

◀ *Attach weatherstripping all round the unit and in the gap formed between the upper and lower sashes.*

ing is important for the removal of condensate.

Step 4. Attach the side wings to the frame.

Step 5. Apply weatherstripping all along the bottom of the lower window sash.

Step 6. Attach additional weatherstripping along the inside portion of the lower window's upper sash to seal the gap between the two windows.

Step 7. Move the lower window down against the frame and secure it with screws.

Step 8. Have someone help you slide the unit into place in the frame. Then secure according to the instructions supplied.

Step 9. Install gaskets around the air conditioner to seal gaps between it and the frame.

Step 10. If the unit requires additional electrical wiring, be sure the wiring is done in conformity with local building codes.

Many room air conditioners can be installed through an outside wall. Since such an installation requires that a hole be cut through the interior wall and on through the exterior wall, however, you may well decide against doing the job yourself or even having it done professionally. On the other hand, if you are experienced in making home repairs, if your interior walls are constructed of drywall,

and if your exterior walls are not made of brick or covered with metal siding, then you can follow these steps to mount your room air conditioner through the wall.

Step 1. Measure the size of the air conditioner housing, and then outline this area on the inside wall.

Step 2. Drill a hole near the top of the area to be removed, and then starting at the hole, cut away a small portion of the wallboard.

Step 3. Check to see what type of insulation is in the wall. If it's the blanket, batt, or rigid type, go ahead and cut away the entire area of wallboard. If the insulation is loose fill, however, cut a slit along the top line of the intended opening large enough to accept a 1x4-inch board sized to fit between the studs within the opening. Work the board into the slit until it meets the outer wall, and then drive several 60d nails into the wall under the 1x4 to act as braces for it. The board will prevent the loose fill above the opening from falling out. When you finish removing the remaining wallboard, you can push the board up even with the top of the opening and toenail it in place. You will need to insert similar stop pieces along the sides, adding more loose fill as indicated before fastening these boards.

Step 4. Drill holes through to the outside wall at the four corners of the opening, and then outline the area to be removed from the exterior.

Step 5. Remove the siding or shingles, the felt below, and the sheathing behind the felt. You can cut the nails that hold the siding with either a hacksaw blade (not in the frame) or with a special tool called a shingle nail remover, but be very careful not to damage the exterior facing. Once the first covering is off, slit the felt with a utility knife and remove it. The wood sheathing can easily be sawed away.

Step 6. Saw out any studs within the opening.

Step 7. Cut and install filler studs between the walls all around the opening. You can use 2x4's or 1x4's or you can even nail the air conditioner frame in place inside the opening. Be sure to seal the opening with caulking or weatherstripping.

Step 8. Remove the housing from the air conditioning unit and mount it inside the opening. The housing should slant down toward the outside to carry away condensate, and it should be caulked properly.

Step 9. Prepare and install a frame made from decorative molding to give a finished look to the inside opening around the unit.

Step 10. Insert the air conditioner in its frame and secure it.

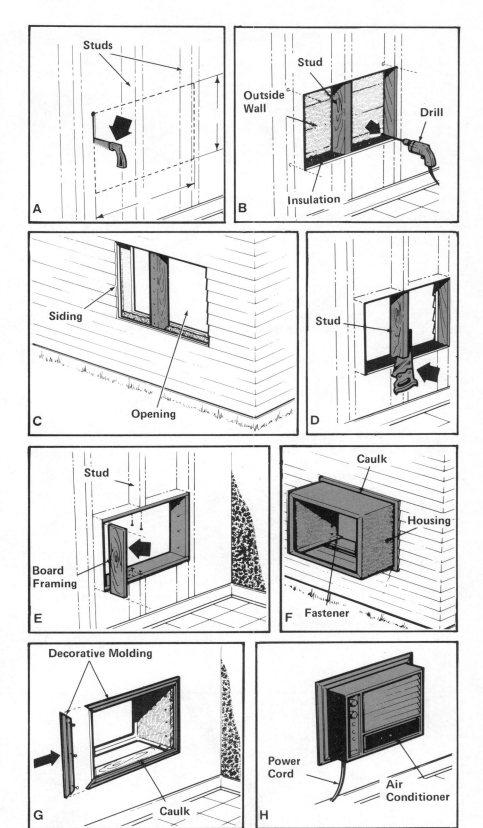

To install a room air conditioner in a wall, (A) measure and cut out the area on the inside wall; (B) drill holes through to the outside wall at the four corners of the opening; (C) remove the exterior wall from the area outlined; (D) saw out any studs within the area; (E) cut and install filler studs between the walls all around the opening; (F) mount the air conditioner housing inside the opening and caulk all around it; (G) install an interior frame made of decorative molding; (H) insert the air conditioner in its frame, secure it, and plug it into a convenient wall outlet.

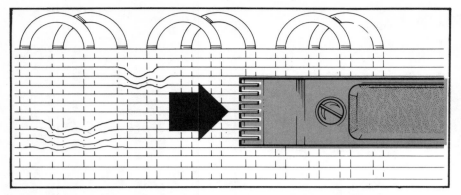

The fins around the condenser coils must be kept clean and straight. If you find that the fins are crimped or bent, straighten them with a fin comb.

Step 11. If needed for cosmetic purposes, you can place a decorative molding frame outside, although usually a plain wooden frame painted to match the house trim will suffice.

Step 12. Check the installation for air leaks, and caulk or patch as necessary.

Maintaining And Repairing Room Air Conditioners

ALTHOUGH REPAIRING a room air conditioner can get complicated, routine maintenance is quite simple. Just follow these steps.

Step 1. Lubricate the fan motor according to the instructions in your owner's manual.

Step 2. Clean or replace the air filter at the beginning of each summer season.

Step 3. Clean and inspect the coil fins. If you discover that the fins are crimped or bent, straighten them with a fin comb.

Step 4. Vacuum the condenser coils and evaporator coils to rid them of dust and dirt.

Step 5. Clean the drain tube that carries off condensate.

When preventive maintenance is not enough and the room air conditioner either fails to function at all or functions unsatisfactorily, the following procedures should get it going again.

Step 1. If nothing happens when you turn the unit on, check the fuse or circuit breaker and replace or reset as necessary. It is possible, of course, that no current is reaching the unit due to a faulty plug and/or cord. Check

and replace any defective electrical parts.

Step 2. When the air conditioner blows warm air, either the temperature control knob (which is actually a thermostat) is faulty or the compressor has gone bad. Check the knob with a continuity tester and replace it if faulty. If the knob is still working, call in a professional repairman to install a new compressor.

Step 3. When the air conditioner blows cool air but not enough of it, check the fan first. You may well find that the bearings are in need of lubrication or that the fan motor requires replacing. Insufficient air flow is also a common result of a lack of routine maintenance; clean or replace the filter, straighten and clean bent fins, and vacuum the condenser and evaporator coils. The flow of air might be hampered, however, by the evaporator icing over; check the blower and make sure that the set screw is tight on the shaft.

Step 4. A compressor that cuts in and out spasmodically may indicate merely that the condenser coils need vacuuming or that some obstruction (like drapes or furniture) must be moved out of the way. It may also indicate that the sensor bulb is out of position; the bulb should be in the return air flow but not touching any of the coils.

Step 5. If your air conditioner seems exceptionally noisy, check

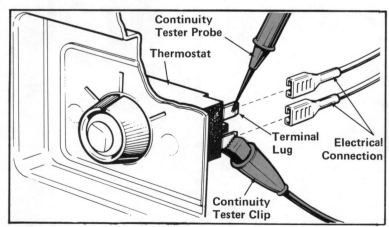

The temperature control knob on a room air conditioner is actually a thermostat. It can be checked with a continuity tester.

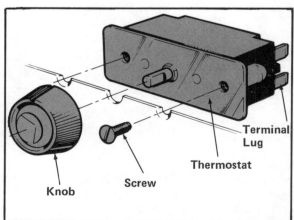

If the temperature control knob is causing the room air conditioner to blow warm air, remove the faulty control and install a new one.

the fan blades and the motor bearings. The blades may require straightening and/or cleaning, and the motor bearings may require lubrication. The noise may also be due to the air conditioner being loose in its mountings; tighten it down or install additional supports as required. Also make sure that the set screw holding the blower to the shaft is tight.

Step 6. Water dripping inside your house from the air conditioner can usually be corrected by repositioning the unit so that it tilts toward the outside or by cleaning the drain tube. If the excess moisture is caused by condensation (a common problem in humid areas), install a drain pan made especially for this purpose.

Ventilation

SINCE HOT AIR rises, intense heat can collect in the attic during the summer unless the attic is properly ventilated. The rising hot air can also carry a good deal of moisture which must be removed as well. Although soffit vents coupled with adequate gable louvers provide a natural way for the attic to be ventilated, an attic fan represents an even better solution. The attic fan draws outside air through open windows to replace the hot air that has risen from the house.

A roof ventilator fan achieves the same objective, but it does so by drawing air up through the soffit vents to replace the hot air it dispels.

Power vents with both a thermostat and a humidistat can be very effective since they come on only when the temperature and/or the humidity reaches a predetermined point—a much more energy-efficient system than a constantly running attic fan.

Air Conditioning

Materials and Supplies

Central Air Conditioners

Flexhermetic II condensing units from Fedders boast high efficiency and low sound levels. The units range in size from 25,000 to 61,000 BTU's and from 8.0 to 8.6 in energy efficiency ratios (EER). Their bristle-fin condenser coils offer improved reliability by reducing the number of joints compared to conventional plate-fin coils. (1)

The Economizer, designed and built by Stiles Corporation, can produce air conditioning energy savings of 20 to 40 percent, depending on geographical location. The system consists of dampers and controls to sense

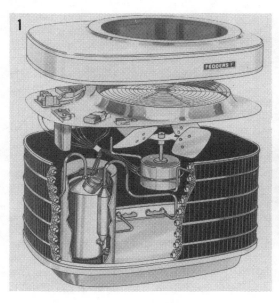

(1) Fedders Flexhermetic II. (2) Stiles Corporation Economizer.

outdoor air temperatures. When the outdoor temperature falls below the indoor temperature, the system automatically shuts off the air conditioner and opens a damper in the roof to allow cool air to enter the house. The air conditioner blower is then used to exhaust inside air and bring in outside air. (2)

Room Air Conditioners

General Electric's Model AGDE-910FC is a 10,000 BTU unit with an energy efficiency ratio of 11.6; it is one of the most efficient room air conditioners available. One of the unit's special features is the Dirt Alert dirty filter indicator. When the air filter gets dirty, the indicator ball rises into view to alert the owner that the filter needs cleaning. A clean filter, of course, keeps the unit operating efficiently. In addition, this GE Superthrust air conditioner has a power saver switch that cycles the fan off and on with the compressor, saving a good deal of energy in the process. (3)

The Friedrich Air Conditioning & Refrigeration Company is clearly concerned about offering high efficiency in room air conditioners. The firm has a full line of units with energy efficiency ratios (EER) higher than 7.5. Of special note is the compact 115

volt model with a capacity of 5200 BTU's and an EER of 9.1. This compact member of Friedrich's Value Series is popular for bedrooms and small living areas.

Covers And Weatherstripping

As expensive as it is to heat a home during the winter, many people compound their heating costs by allowing cold air to enter their homes through the window air conditioner. An inexpensive easy-to-install plastic cover seals the air conditioner effectively. Allied Plastics makes an extra heavy gauge (5 mil) polyethylene cover to fit over the outside portions of most window units. To seal the window unit even more securely, tape pieces of insulation to the outside of the air conditioner before attaching the plastic cover.

Frost-King makes both indoor and outdoor covers for a room air conditioner. The outdoor unit is made of heavy vinyl and comes in two sizes with adjustable straps. The indoor cover is heavy-gauge polyethylene. Although many people attach both, an inside cover is especially valuable to high-rise apartment dwellers who cannot safely seal their air conditioners from the outside. (4)

If you have an exhaust fan that lets warm air out during the winter and air conditioned air out during the summer—or if it allows outside air to enter the house—you should consider buying a Crest Magnetized Exhaust Fan Cover. Designed to fit over standard eight-inch round exhaust vents, the cover snaps easily in place over the metal vent and comes off just as easily when you need to run the fan.

Room air conditioners, if not properly sealed, can really let the cold air out and the hot air in. Air can escape between the unit and the window frame of the raised window and in the space between the two sashes in double-hung windows. Mortell Windowseal is designed to seal all the gaps around the unit and the window. One package of Windowseal all-season foam will seal one window air conditioner. Self-adhering tabs to keep the foam weatherstripping in place are included.

Air Conditioner Accessories

The moisture that your air conditioner takes out of the air drips down into a pan and is carried outside. Sometimes, however, a fungus can grow in the pan or the drain tube and clog up the system. This usually causes an overflow that can damage the house and furnishings. A cup of

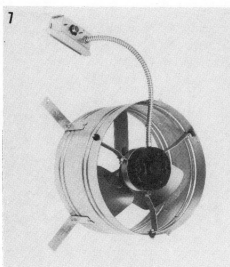

(3) GE Model AGDE-910FC. (4) Frost-King outdoor air conditioner cover. (5) Kool-O-Matic attic ventilator. (6) Midget Louver. (7) Broan attic ventilator. (8) AMF/Paragon air conditioner timer.

laundry bleach poured into the pan at regular intervals will retard the growth of the fungus, but there is another way for people who don't wish to be bothered with such chores. The Anti-Clog Unit from CDC Chemical Corporation sits in the pan and releases a chemical to fight fungus growth; it even features a patented Time Capsule Cap to release the chemical automatically only when needed. The CDC Anti-Clog Unit comes in two sizes for use with central air and with room air conditioners.

A room air conditioner timer, such as the one from AMF/Paragon Electric Company, turns the cooling unit off at any predetermined time and turns it back on at a later time, allowing you to go to work and return to comfort without running the air conditioner all day. The timer needs no special installation. Just plug the timer into an electrical outlet, and then plug the room air conditioner into the timer. (8)

An excellent little book that explains how central air conditioning works, how to keep yours working and what to do when it stops is titled *Is Your Air Conditioner Blowing Cold Or Blowing Money?* Written and published by Gene Elwell, it is available by mail order for $3.00 postpaid to Gene Elwell, 818 Fontana Avenue, Richardson, Texas 75080. The hints it provides on efficient air conditioner operation will more than pay for the book, to say nothing of how much it can save in repair costs.

Attic Ventilators

H-C Products Company has found a way to let hot attic air escape along the roof ridge. Vent-A-Ridge consists of aluminum alloy sections that allow hot air to escape while making it impossible for rain or snow to get in. Since no fan is involved, the system consumes no energy. Installation is a do-it-yourself project that can be completed in a day's time.

Thermostatically controlled power vents from Leslie-Locke —equipped with optional humidistats—can be quite valuable in saving energy. By removing hot air and excess humidity during the summer, these ventilators ease the load on air conditioning systems. Available in roof-mounted and in sidewall or gable-mounted versions, Leslie-Locke units feature an adjustable automatic thermostat, permanently lubricated motor, insect screening, cushioned motor mounts and reliable safety devices. In addition, they all carry a lifetime limited warranty.

A Kool-O-Matic attic ventilator can be a great help in ventilating your attic and keeping your cooling costs down. When the temperature in the attic reaches 100 degrees, the unit's thermostat turns the ventilator on and keeps it on until the temperature drops to 85 degrees. The Kool-O-Matic's humidistat turns the unit on when the humidity reaches 90 percent. The Kool-O-Matic Ventilator can be installed on a sloping roof, flat roof, or vertically on gable ends. (5)

In order to ventilate your home as much as possible, you can add small vents in overhangs, gable ends, walls and many other places. The Midget Louver Company makes small louvers that can be installed easily. Just drill an appropriate sized hole, push the Midget Louver in place, and tighten the special fasteners. The louvers are made in several sizes and styles. (6)

Broan Manufacturing Company makes an energy-saving attic ventilator designed to be gable-mounted behind a louvered panel. By cooling the attic, the thermostatically controlled ventilator reduces the running time of air conditioners during hot weather. The reduced running time can save energy consumption by as much as 30 percent. (7)

Appliance Efficiency

ALTHOUGH THE figures vary with the type of home and family, appliances generally account for about 20 percent of the home's total energy consumption. While the amount of energy saved through appliance efficiency is not nearly equal to that gained through weatherproofing the home, you can achieve noticeable cuts in your utility bills by being energy conscious when you shop for new appliances and when you care for the ones you now own.

Refrigerators And Freezers

THE REFRIGERATOR/freezer (frequently combined in a single unit) is usually the single greatest energy consumer of all the appliances in a typical home. Since improper maintenance can cause wasted energy on top of the unit's normally large energy usage, wise homeowners keep their refrigerator/freezers in good working order.

Here are the steps to follow.

Step 1. Cleaning the condenser coils about four times a year is an absolute must. Use a tank-type vacuum cleaner with a brush attachment to remove efficiency-reducing dust on the coils.

Step 2. Many refrigerators have a pan underneath to catch condensate water. Check the pan each time you vacuum the condenser coils, and clean it as necessary. If it is not cleaned regularly, the pan will collect particles of food and cause odors.

Step 3. If your freezer is not self defrosting, watch the frost build-up. When the frost level approaches ¼ of an inch, defrost your freezer. Otherwise, the layer of frost will act as an insulator and cut down on the freezer's efficiency.

In addition, you should check the temperature in both the refrigerator and the freezer to be sure that the controls are set at the most economical positions. The temperature in the freezer should be between zero and five degrees. Place a thermometer between packages that have been in the freezer for at least 24 hours. If the reading is below zero or above five degrees, reset the freezer thermostat up or down accordingly.

The temperature in the refrigerator should be between 34 and 37 degrees. Put the thermometer in the refrigerator long enough to get an accurate reading, and then set the refrigerator temperature control up or down to bring

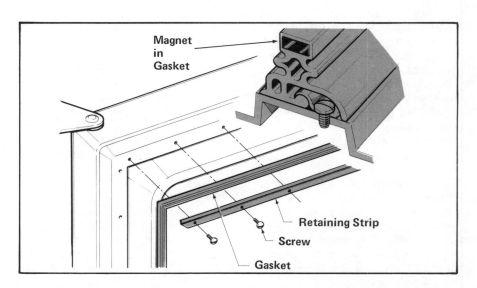

The refrigerator gasket, held to the door by screws through a retaining strip (left), usually has magnets inside to create a tight door seal (right).

it within the proper range.

Check for a faulty door gasket by closing the door on a dollar bill. With the door closed, tug at the dollar. If the bill comes out without much resistance, the refrigerator door is not sealing properly. Perform this dollar test at several other places around the door.

A faulty gasket or loose hinges can prevent a tight seal. Most gaskets have magnets inside them to hold the door closed. If you need a new gasket, you can generally replace the old one yourself. It is held to the door by screws that go through a metal retainer strip, an inner door facing, and then into the door. To get at the screws, peel back the old gasket, remove the screws in the top half, and install the new gasket across the top. This prevents the entire inner door facing and shelves from coming off, which would happen if you removed all the screws at once. When the new gasket is in place, make sure that it makes contact all the way around. Smooth it out with your hands, and rub out big kinks with hot towels.

Ice maker problems occur when the strainer in the water line entry gets clogged. Cut off the water supply and unscrew the hose type connector on the back of the box. Then clean the strainer with an old toothbrush.

Automatic Dishwashers

A GOOD dishwasher, kept in proper running condition, uses less water than hand-washing the same amount of dishes would require. Of course, it does consume electricity, but you can keep your dishwasher from becoming an energy waster by using it only when it is full of dirty dishes and by keeping it in good operating condition.

Refrigerator/Freezer Energy-Saving Tips

1. Avoid opening the refrigerator and/or freezer doors. Plan ahead and remove as many items as you'll need at one opening.

2. Don't set the temperature dials too low. Experiment to arrive at the highest freezer setting that will keep ice cream firm but not rock hard. The refrigerator temperature should stay within the range of 34 to 37 degrees.

3. Keep your freezer full; frozen foods help retain the cold. Do not, however, put in large amounts of unfrozen food to be frozen all at once.

4. Leave enough space in the refrigerator for air to circulate.

5. If you have a frost-free freezer, keep all liquids tightly covered. Uncovered liquids evaporate and cause a frost-free system to work harder.

6. Let hot dishes cool before putting them in the refrigerator.

7. When you go on a vacation, empty the refrigerator as much as possible and raise its temperature setting slightly.

8. Never install a refrigerator or freezer near any heat-producing appliances, and always allow for the proper air circulation around the unit.

9. Keep your refrigerator and freezer clean and in proper working condition.

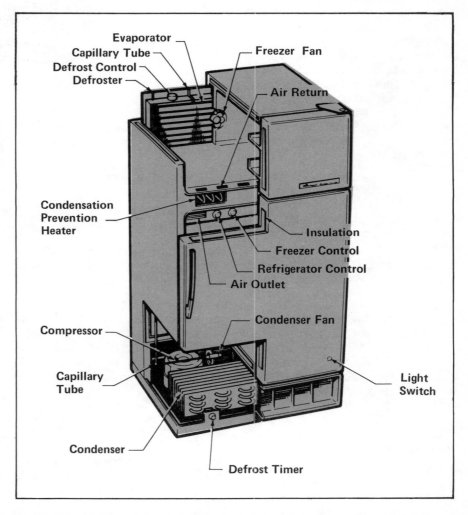

Evaporator
Capillary Tube
Defrost Control
Defroster
Freezer Fan
Air Return
Condensation Prevention Heater
Insulation
Freezer Control
Refrigerator Control
Air Outlet
Condenser Fan
Compressor
Capillary Tube
Light Switch
Condenser
Defrost Timer

The refrigerator/freezer is usually the single greatest energy consumer of all the appliances in a typical home. It must be kept in good working order.

Dishwasher Energy-Saving Tips

1. Run the unit only when it contains a full load.

2. Don't pre-rinse, but do scrape away all food particles from the dish surfaces. If dishes have been left all day waiting for a full load, rinse away any hardened bits of food.

3. Follow the manufacturer's suggestions on loading. This procedure lets the soiled parts face the washing action and is very important for maximum efficiency.

4. Follow the manufacturer's suggestions on keeping the filter screen free of food particles.

5. By shutting off the dishwasher just before it goes into the drying cycle, you'll save a considerable amount of energy. Many of the newer dishwashers feature a shutoff setting to perform this function automatically. After the dishwasher shuts off, open the door and let the dishes air dry. During the summer, turn on an exhaust fan to remove the steam from your kitchen.

6. Be sure to use the proper amount of detergent. Too much or too little or even the wrong kind of detergent reduces your dishwasher's efficiency.

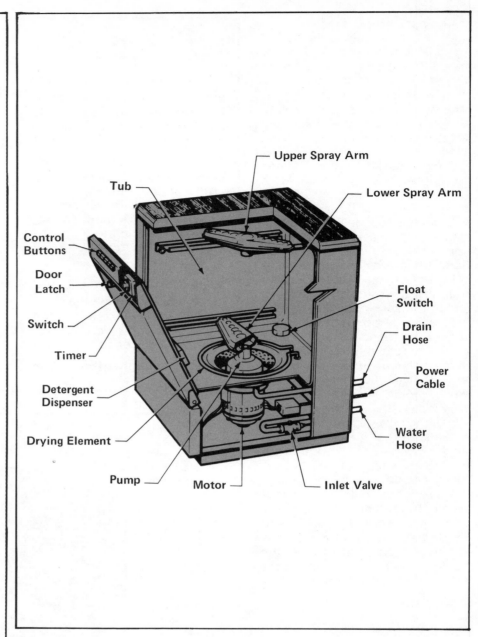

A properly functioning automatic dishwasher will not consume a great deal of electricity and will clean a full load using less water than hand washing.

One very complicated component in a dishwasher is the timer. Fortunately, you do not have to know how this gadget works to ascertain that it needs to be replaced. Here are the steps in replacing a dishwasher timer.

Step 1. Turn off the electric power to the unit.

Step 2. Remove the panel from the dishwasher to expose the timer. Most knobs pull straight off.

Step 3. Hold the new timer next to the old one, and change the wires one at a time from the old unit to the new one. That way, you will not get the wires crossed up.

Step 4. To disengage the wires, pull them straight off, and then slide them on the new terminals.

Step 5. Now remove the screws holding the old timer, pull it out, and replace it with the new unit.

The spray arm (or arms) and the screen can get clogged with particles. Unscrew the knurled nut above the arms to remove them, and then clean the arms in the sink with a brush.

The pump can also become clogged. Consult your owner's manual for the procedure to get to and clean the pump. Be sure to keep track of where everything goes (a rough sketch can be a great help) as you remove parts so that you can put them back in their proper places. It is best to replace any O rings you remove with new ones even though the old ones may look all right.

Ovens And Stoves

ENERGY CONSUMED in cooking food is nearly equal to that used in cooling or freezing it. Whether your stove is gas or electric, keeping it clean and in proper working order will allow you to get the most from your unit.

In a gas range, the heat for both the surface burners and the oven emanates from an open flame. The fuel may be either natural gas or one of several types of bottled gas, but the operation in either case is the same. The gas burner operates by combining the fuel from the supply line with the correct proportion of air for thorough and clean burning.

The fuel enters the burner assembly through an orifice, which is simply an opening sized to provide.the proper amount of gas flow. The gas stream pulls air in through an open shutter behind the orifice, and as the mixture flows on through the burner tube to the burner itself, the air and gas mix thoroughly. By the time it reaches the burner, the air/gas mixture is ready to meet the pilot or the burner flame. Burning then occurs with an odorless and soot-free flame.

Air must be mixed with the fuel supply in the proper proportion to provide a clean-burning and efficient flame. You can judge the quality of the air/fuel mixture by observing the flame. If the flame is yellow, you know that there is insufficient air; if the flame tends to pull away from the burner, you know that there is too much air.

You can regulate the quantity of air by adjusting the shutter, which is located at the point where the burner meets a pipe near the front of the range. Loosen the screw and close the shutter until the flame turns yellow. Then open the shutter slowly until the yellow tips of the flame just disappear. You should see a distinct cone-shaped flame with soft blue tips. The flame is hottest at the points of the tips, and coolest at their base. When the shutter is set to render the correct flame, retighten the screw.

The heat output from each surface burner is usually regulated by a separate control valve which opens and closes an internal passageway to allow more or less gas to flow though the orifice. The amount of gas that flows through the orifice automatically adjusts the air mixture; a smaller amount of gas flow allows a smaller quantity of air to enter the air shutter.

The oven thermostat may need adjustment. Here is how it works. A sensing tube runs from the thermostat into the oven chamber. Inside the tube, a liquid responds to heat changes by expanding or contracting, an action which in turn opens or closes a switch within the body of the thermostat.

The switch controls an electrical coil, called a solenoid, located on the gas line. The solenoid, consisting of coils or wire that are wound around a central armature or plunger, concentrates the magnetic force generated by the electricity flowing through the wires. When the coil is energized, the plunger moves,

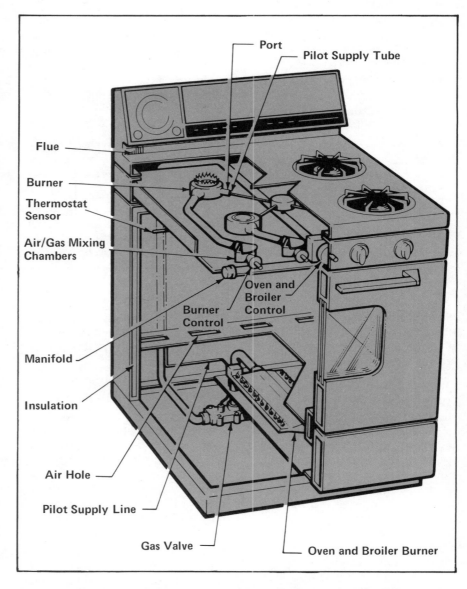

In a gas range, the heat for both the surface burners and the oven burners comes from an open flame. The burners combine the fuel from the supply line (either natural or bottled gas) with air for clean and complete combustion.

**Oven And Stove
Energy-Saving Tips**

1. Cook with as little water as possible since small amounts heat quicker.

2. Put a lid on the pot or pan you're using. Water boils faster when covered, saving up to 20 percent of the energy that would otherwise be consumed.

3. Fit the pot to the burner. A small pot on a large element is an energy waster. Adjust the element setting (if possible) or the gas flame to fit the size of the pot you're using.

4. Turn down the heat as soon as the liquid you're heating reaches the boiling stage. Adjust the setting so as just to keep the contents boiling; a higher setting is an energy waster.

5. Thaw frozen foods before cooking them.

6. Keep the bottoms of pans clean. Shiny pans are particularly efficient in electric cooking, but a layer of soot will decrease heating efficiency no matter what kind of stove you have.

7. If you have a choice, use the stove-top burners or elements for foods that cook quickly and the oven for those that will take a long time. The stove-top units consume less energy, while the oven retains heat for prolonged periods.

8. Cook with a microwave oven whenever possible.

9. Don't turn on an element or burner until the pan is on the stove.

10. Follow the recommendations of the stove manufacturer as to the type of cookware you should use. In most cases, copper and stainless steel cookware require lower heat settings than does aluminum.

11. With an electric burner, the cooking process will often continue for as much as five minutes after you turn it off. Even with gas, some cooking continues after heat is shut off, but only for about a minute.

12. When you plan to use your broiler or oven, try to cook as many things as possible at one time.

13. Do not preheat the oven unless absolutely necessary. If you must preheat, be ready to put the food in just as soon as the oven reaches the desired temperature.

14. Never preheat when broiling.

15. Try lower oven temperature settings when you bake in glass or ceramic dishes.

16. Never use your oven or stove to heat the kitchen.

17. Keep your inspections of food cooking inside the oven to a minimum. Every time you open the oven to take a look at what's cooking, you waste a tremendous amount of heat. During the summer, of course, that extra heat puts a strain on your air conditioner.

18. Put a self-cleaning oven into the cleaning stage right after cooking since the oven will already be far along toward the high temperature needed for cleaning.

19. If you have a double oven, use the smaller one whenever possible.

20. Use small cooking appliances (e.g., toasters, toaster ovens, crockpots, electric frypans, etc.) instead of your range whenever practical. These small units consume less energy and throw less heat into your kitchen.

---

opening or closing the gas line to the oven burner as required to maintain a specific temperature setting.

Some oven thermostats operate directly on the gas line, moving a bellows arrangement to open or close a disk which, in turn, permits or blocks the flow of gas through the line. Some of these valves modulate—that is, they open or close the gas line gradually rather than instantaneously. A modulating valve reduces any temperature overshooting, and it helps maintain a more constant temperature level.

You can adjust the thermostat of a gas range if the temperature within the oven varies by more than 25 degrees from the setting on the knob. Here's how to do it.

Step 1. Check the oven temperature by placing an accurate thermometer in the center of the oven cavity.

Step 2. Allow the oven burner to stabilize for 30 minutes.

Step 3. If you find a great disparity between the thermometer reading and the knob setting, remove the thermostat knob and look for an adjustment screw. You may find the screw within a hollow thermostat shaft, or you may find it on a movable scale at the rear of the knob skirt.

Step 4. Move the screw or scale until the setting on the knob corresponds with the actual oven temperature.

Gas ranges require little in the way of service. Their electrical components—timer, thermostat, and valve—are subject to failure, but they rarely do fail. Here are some helpful maintenance tips.

1. Before you perform any sort of service procedure, be sure that you shut off the gas supply and unplug the range.

2. You can disassemble many burners, soak them in hot soapy water, and then brush them with an old toothbrush to remove food particles.

3. If you notice that a burner is starting to clog, you can prevent the clog from worsening by cleaning the blocked orifice with a wooden toothpick.

4. If you see that the heat output from a particular burner is reduced below its normal level, use only a soft object (like a toothpick) for cleaning. Since metallic objects can enlarge the orifice openings, they should not be used for cleaning.

5. Should a pilot light become clogged, it might be necessary to unscrew the orifice tip itself and clean the orifice from the inside. The opening in the orifice is generally too small to be cleaned from the outside.

6. If a burner hesitates when you turn on the gas supply, check to be sure that the pilot flame is adjusted correctly and that the connecting tubes from the pilot to burner are in place.

If you smell a gas leak, be sure to call a technician immediately to inspect the range, and do not use the range until it has been examined thoroughly. Open the windows to provide plenty of ventilation, and extinguish any open flames. Natural gas itself has no odor, but the gas company adds an artificial odor to help you detect leaks in gas lines. A leak indicates a situation that is potentially very hazardous. Should you ever smell gas in your home, call professional service personnel immediately.

Cooking with electricity (excluding a microwave oven) is more costly than with gas. It is even more important, therefore, to keep your electric range operating with maximum energy efficiency.

Most electric ranges employ a sheath-type enclosed nichrome heating element to provide a controlled amount of heat to the cooking surface and to the oven cavity. The heating elements on the top of the range are shaped to make the maximum amount of contact with the bottom of your pots and pans, which must be flat and in good condition. Otherwise, the pan can produce "hot spots" in the element, and hot spots reduce the life of the element as well as yield poor cooking results.

The cooktop elements—or surface units as they are called—can

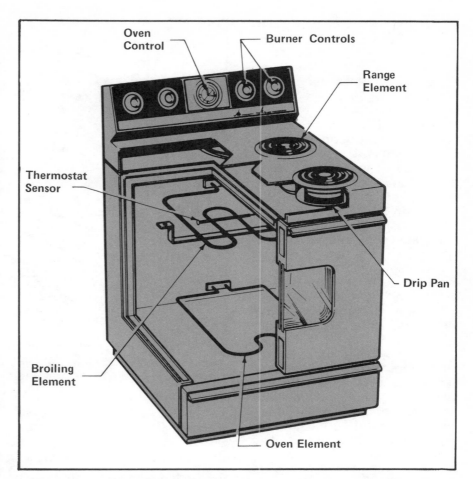

Since an electric range is even more expensive to operate than a comparable gas unit, it is essential that the electric range function with maximum energy efficiency. Watch for burned-out elements and poor wiring connections at both the surface and oven elements. All wiring connections must be tight and free from corrosion for elements to function at peak efficiency.

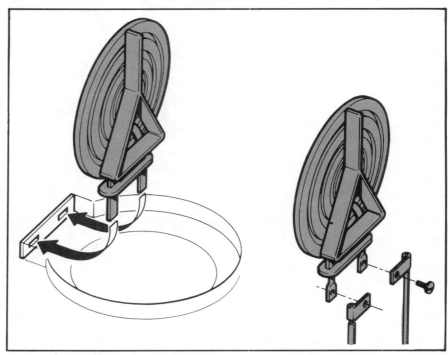

Surface elements can generally be removed quite easily. They either plug into special receptacles at the rear (left), or they are hinged to make their wiring accessible (right). Replace any faulty element promptly.

usually be removed easily. On many newer ranges, in fact, the surface units simply plug into a special receptacle at the rear, while on some other ranges, the units are hinged to make their wiring accessible.

Some electric ranges include a thermostat control on one of the surface units. Usually a solid state type of resistor, this thermostat changes resistance according to the amount of heat you set on the control. The thermostat control has a heat sensor that rests lightly against the pan bottom, and the control adjusts the element to maintain an exact temperature level at the bottom of the pan. It is absolutely essential, however, that your pan have a flat surface that meets the sensor squarely.

Some new ranges have flat ceramic tops over the elements; heat is transferred from the element through the ceramic material to your pots and pans. Sometimes, separate thermostatically controlled switches are used to provide controlled heat to each section of the flat-top range. Be sure to follow the manufacturer's recommendations for cleaning these ranges, and always treat the ceramic surfaces with great care.

Many surface-element problems are related to a burned-out element or to a poor wiring connection either at the element receptacle or where the wire is attached to the element. You can repair any of these connections easily, but be sure to turn off the power supply and unplug the range before attempting to fix any part of your range.

It is possible for an element's inner coil to be grounded to its sheathing. If this happens, the coil can actually burn a hole through the outer surface of the element. Sometimes the hole is visible, but if it is not, test the element according to the following procedure.

Step 1. Use a continuity tester, and if you discover a fault, obtain a replacement element from an appliance parts distributor.

Step 2. Record the color of each wire and the terminal to which it

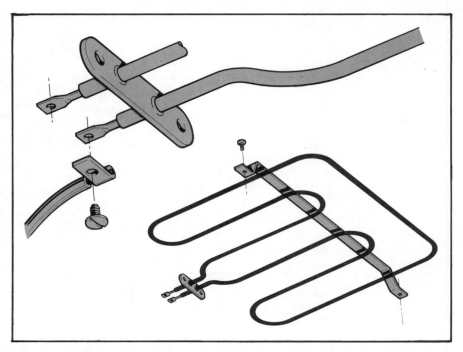

You can remove a faulty oven element by turning off the power, loosening the mounting screws, pulling the unit forward, and disconnecting the wiring.

is connected before you disassemble the faulty element. Rewiring the replacement element should not offer any problems.

Step 3. If a wire should break or burn away from the element or receptacle, install a special high-temperature terminal.

Step 4. Be sure that all the wiring connections are tight, and if any connections appear to be corroded, polish them with a file or sandpaper until they are bright and shiny. Heat from a poor connection is usually the cause of such corrosion, and if you fail to clean it, the connection is sure to malfunction again.

The oven has a capillary-tube thermostat to detect and maintain the proper temperature levels. The thermostat is usually not serviceable; if part of the thermostat fails, the entire control must be replaced. If the thermostat is merely out of calibration, however, you usually can adjust it. Here's how to do it.

Step 1. Place an accurate thermometer as near to the center of the oven as possible, and set the thermostat to a temperature in the medium range.

Step 2. Allow the oven to stabi-

lize for approximately 30 minutes before taking a reading.

Step 3. If there is a great disparity between the thermometer reading and the thermostat setting, remove the thermostat knob and look for a calibration screw; you may find the screw within a hollow shaft or on the front of the control.

Step 4. Turn the screw to correct the thermostat setting.

Step 5. In some electric ranges, the skirt on the back of the thermostat knob is adjustable. Loosen the screw and adjust the position of the skirt to get the thermostat knob indicator to the correct temperature.

Step 6. If the thermostat moves out of calibration after you adjust it, install a new thermostat.

If a heating element in an oven fails to operate, and you have verified with the continuity tester or by visual inspection that it is defective, remove it from the oven by loosening the mounting screws at the rear of the oven. When you pull the defective element forward, enough connecting wiring should come with it to allow you to remove the wire from the old element and to attach it to the new element. Naturally, the power must be turned off before you attempt such a repair. In addition, make sure that the wiring connections are bright and shiny before you install the new element.

Many ranges have a timer connected to the oven circuit to turn the oven on and off at preset intervals. Although a convenience feature, the timer, ironically, is frequently the cause of people thinking that their ranges are defective. The timer controls can keep the oven from operating if they are inadvertently moved out of position. Therefore, if your oven fails to function, make certain that the timer is in the manual or normal position.

The outlet and light are usually connected to the 110-volt circuit independent of the other range circuits. Generally, a separate fuse (located beneath the elements) or a circuit breaker (on the control panel) protects the outlet and light circuit. If the outlet or light ever fails to operate, check the condition of the fuse or circuit breaker.

If the entire range fails to operate—or if it operates only at a low temperature—chances are that one or both of the main fuses in your home's electrical circuit to the range have blown. If they are cartridge-type fuses, you must check them with a continuity tester or replace them with new ones. If the range circuit is protected by a circuit breaker, simply reset it to restore the range's power supply. Of course, you should try to determine why the fuse blew or the circuit breaker tripped.

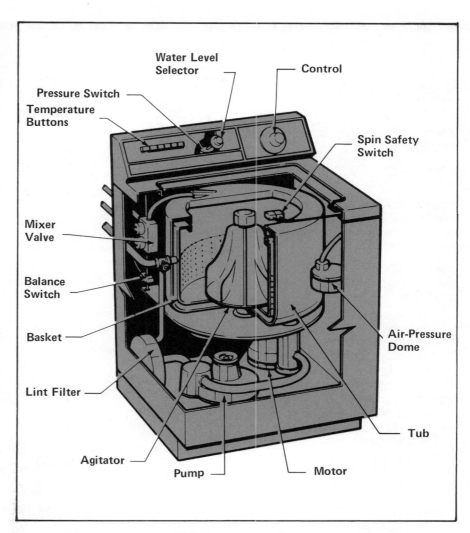

A clothes washer that is not maintained properly can waste a good deal of both electricity and hot water.

cared for, it can waste both.

There are many complicated systems within a modern clothes washer. Many of these parts require special equipment to track down problems, but there are several common problems that the average homeowner can handle by him- or herself. Just make sure that you shut off the current to the unit before making any repairs.

A slow-filling washer can be the result of clogging in the tiny

Clothes Washers

THE CLOTHES WASHER consumes both hot water and electricity, and if it is not properly

Clothes Washer Energy-Saving Tips

1. If your washer has a water level control, adjust it for the least amount of water needed to do the job. Do not, however, overload the level selected. Overloading results in the unit having to work beyond its normal capacity.

2. Clean the lint filter after each load.

3. Wash clothes of similar fabrics together. Use the shortest washing cycle for the type of material involved, and use the coldest water setting that will get the job done right.

4. Watch the amount of detergent you add. Oversudsing can overwork the machine.

screens in the line where both hot and cold water enter the machine. Here's how to clean them.

Step 1. Turn off the faucets and unscrew the hoses that are attached to the back of the washer.

Step 2. The screens may have to be pried out.

Step 3. Clean the strainers with a stiff brush.

Step 4. In reattaching the hoses, put them on hand tight, and then add a quarter turn more with a pair of pliers.

Step 5. Check to be sure that there are no leaks after replacing the hoses.

The water line hoses as well as the hoses within the actual unit can develop leaks. Fortunately, these hoses are easy to replace. They are held in place by spring-type hose clamps that must be squeezed with pliers to loosen. Be ready to catch water from the hoses with a towel.

Anytime you work on a washer, make sure to check the unit with your level after you put the washer back in place. A washer that is not level must work much harder, wears out sooner, is much noisier, and fails to do its cleaning job as well as it should.

Clothes Dryers

THE CLOTHES DRYER consists of a cabinet surrounding a motor-driven drum. Clothing placed inside the drum tumbles through air which has been warmed by a heater and pulled into and through the drum by an exhaust fan. These three basic components—heat, air flow, and tumbling action—must be present for a clothes dryer to operate properly.

Actually, the clothes dryer is a very simple appliance. When it is operating properly, it can dry your clothing quickly and safely, but the appliance must be kept in proper operating condition and it must be used correctly.

The motor that drives the drum is usually located at the base of the dryer. You can get at it by removing either the front panel or the rear service panel, but be sure to unplug the dryer before attempting any motor service or repairs. In the case of a gas dryer, be sure that the gas line is turned off before you open up the appliance or pull it away from the wall.

The motor drives a belt, which in most newer dryers completely surrounds the outside of the drum. Many of these belts have an odd appearance—almost flat like a rubber band. If you look at the belt closely, however, you can see that it has a number of grooves on one side to give the belt greater gripping power.

Near the point where the belt is attached to the small motor pulley, a spring-loaded wheel (called an idler) maintains tension on the belt as the drum turns. Since the drum itself acts as a very large pulley and the motor has a very small one, tremendous speed reduction is obtained. Most drums rotate at around 50 rpm. Any faster and centrifugal force would tend to hold the clothing against the side of the drum rather than allowing it to tumble.

The dryer's heat source can be either an electric heating element or a gas burner. In either case, the heat source is usually located within a box that has both an inlet and an outlet. Air flows in through the box where it is heated, and then it is blown into the drum by a fan. Since the drum is a sealed container when the dryer door is closed, the exhaust fan must pull air through the opening and into the heater box to replace that expelled by the fan. The heated air flows into the drum, where it absorbs moisture from the clothing, and it is then exhausted to the outside of your home through the vent.

The heat level is quite important. Heat must be controlled at the proper temperature for the type of clothing that is in the dryer. Most dryers have one thermostatic control set to operate around 145 degrees F., but some dryers offer an adjustable thermostat which you can set according to the nature of the load you are drying. The adjustable thermostat can range from 120 degrees for delicate clothing to 155 degrees for heavy cottons and linens.

You can check your dryer's temperature by inserting a candy or meat thermometer in the exhaust vent at the point where the vent is attached to the dryer. The machine's temperature sensors are usually located within the exhaust duct, just inside the dryer.

There is another thermostat in your dryer, but it is there for safety purposes only. The safety thermostat shuts off the heating element or burner when the temperature in the dryer exceeds 200 degrees. Normally, your dryer will never even approach 200 degrees; it gets that hot only in cases where the thermostat sticks, the heating element becomes grounded, or the air flow through the dryer is blocked.

The following list describes some of the things you should and should not do to your clothes dryer.

1. Most dryers that were built within the past few years have sealed-for-life lubrication. If you ever disassemble the dryer for any reason, however, you must be sure that all bearing surfaces are lubricated; use the manufacturer's recommended lubricant. One important spot to check is around the front edge. The drum of many modern dryers is supported on rollers at the rear, while in front the flange of the door opening serves as a bearing surface. If the surface appears to be dry, lubricate it according to the manufacturer's instructions.

2. One of the most important preventative maintenance tasks you can do for your dryer is to clean the lint filter prior to drying every load. If the filter becomes completely clogged, some lint can escape and create jamming problems elsewhere in the dryer. Even a partial blockage reduces the dryer's efficiency and limits its capabilities. Most importantly, though, a clogged dryer can be a fire hazard. Lint from many fabrics—particularly synthetics—is highly combustible.

3. The vent is designed to carry heat and moisture away from the dryer to the outside of the house. While the vent may seem to waste a great deal of heat which could be put to use, remember that the warm air is heavy with moisture after it passes through the dryer. Were this air to be recirculated through the dryer, the appliance's efficiency would suffer greatly, since the air simply could not hold much additional moisture. Moreover, the same air is circulated through the dryer motor to help cool it. The hotter the air, the hotter the motor.

4. Although it is very tempting to place large quantities of clothing in the big drum, remember that the clothing needs a great deal of space for tumbling. Never dry more than a single washer load in a single dryer load, and never try to bake your clothes completely dry. Most clothes should be allowed to retain a slight amount of moisture.

5. Once a year unplug the dryer or turn off the gas supply, remove the service panel, and vacuum away any lint or dust in the vicinity of the motor. Regular cleaning keeps lint away from the bearings, and it helps maintain clean air passageways. It also reduces the possibility of a fire.

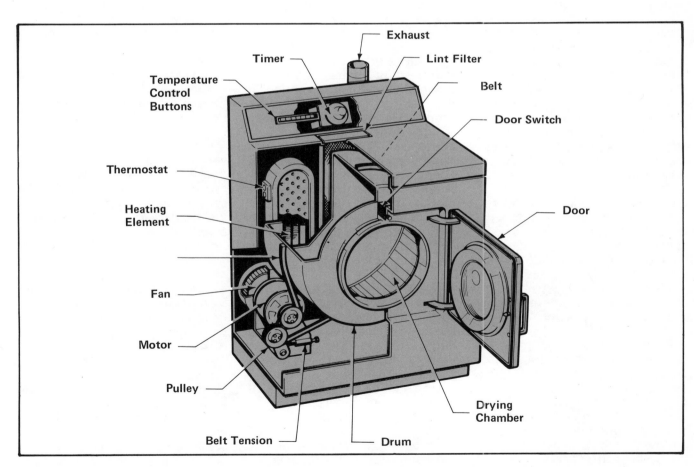

A clothes dryer, either gas or electric, wastes energy when run for partial loads or with excessively high heat settings.

Appliance Efficiency

Materials and Supplies

Refrigerators And Freezers

Hotpoint's CTF18HV is a 17.6 cubic foot no-frost refrigerator with improved energy efficiency compared to other Hotpoint refrigerators of the same capacity. Poured foam insulation, greater condenser surface area, improved door insulation and a low-power condenser fan motor combine to produce a 30 percent reduction in energy consumption compared to previous Hotpoint models. The unit consumes between 89 and 121 kWh per month, depending on the setting of its Energy Saver Switch.

Whirlpool's Model EET171HK, a 17 cubic foot no-frost refrigerator, consumes 94 kWh per month with its power-saving switch on low. The previous model consumed 99 kWh per month, which was quite low for a unit of its size. The increase in efficiency is largely due to an improved urethane foam insulation system. (3)

Admiral Model NS2277, a 21.8 cubic foot side-by-side automatic defrost refrigerator/freezer, consumes 151 kWh of electricity per month. The typical antisweat electric door heaters have been replaced by a condenser loop that warms the refrigerator door and saves energy in the process.

Free-O'-Frost refrigerator/freezers from Amana offer energy savings of 17 to 41 percent compared with competitive models. The units, called "2 Plus 2½," have 2½ inches of foam insulation around the freezer compartment and 2 inches around the fresh food section. By comparison, conventional refrigerators have about 1¾ inches of insulation around both freezer and fresh food sections. Four energy-saving refrigerators are available from Amana in sizes ranging from 14.2 to 16.2 cubic feet.

Power-saving heater control switches on freezers can cut your electric bill. All Whirlpool uprights, except for the 12 cubic foot model, feature such a switch that allows you to turn down the antisweat door heater during periods of low humidity. Whirlpool's Model EEV161F, a 15.9 cubic foot unit, has the switch and is a manual defrost freezer for extra energy savings.

Amana's ESU-17, a 17.1 cubic foot upright freezer, combines 2½ inches of foam insulation in the cabinet (compared to the normal 1¾ inches), a one-piece molded inner liner, and foam/fiberglass insulation in the door to produce major energy savings. The manual defrost unit consumes 75 kWh of electricity per month. The ESU-15, a 15.1 cubic foot model, consumes 68 kWh per month, while the ESU-13—a 13.1 cubic foot freezer—needs 60 kWh per month.

Foamed-in-place insulation makes the 25.3 cubic foot chest freezer from White-Westinghouse an efficient unit despite its large size. The manual defrost unit, Model FC258T, features a built-in lock, exterior defrost drain and a safety signal light. It consumes about 114 kWh of electricity per month.

Frost-King's replacement refrigerator gasket is molded of pure rubber to conform to any contour and to remain pliable for many years. The Frost-King kit contains 17 feet of gasket, and it can come in very handy if you can't obtain a gasket made specifically for your unit.

Dishwashers

Dishwashers that are equipped with controls that allow dishes to dry without running an electric heater can save energy. This White-Westinghouse Model SC650W is such a model. The control cuts off the powered drying system that draws air into the tub, heats it and then fan circulates it to dry the dishes, reducing energy consumption by 30 percent. This top-of-the-line dishwasher is a convertible unit that can be used as a portable or installed as a built-in under a counter.

Magic Chef's eight-cycle convertible (portable or built-in) dishwasher offers two short cycles that can save up to 60 percent of the energy consumed by the regular cycles. Two additional energy-saving features include the full wash/cool dry and full wash/no heat dry options.

The Dishmaster from Manville Manufacturing Corporation can help you clean dishes without working much harder than you would loading an automatic dishwasher. Dishmaster replaces the old faucet unit and performs all of the conventional faucet operations, but it does more. Just feed a tablespoon of detergent into the tank and a diverter sends soapy water or clear as needed out through the brush handle. For a small family with few dishes to wash, the Dishmaster presents an attractive way to conserve water and energy. (1)

Washers And Dryers

The appliance industry's first solid-state electronic touch control washer, Whirlpool's Model LFA 9800, has special options that can help save energy. These energy-saving options include the short wash, extra-short wash and cold wash. The unit also has a water-saving load size selector. Although an expensive machine, the LFA 9800's electronic controls offer simplicity and versatility.

The new generation washers from White-Westinghouse are designed to offer dramatic reductions in water, energy, detergent and bleach consumption. The front-loading, tumble-action washers use 36 percent less water (41 percent less hot water), 35 percent less electricity, and 67 percent less detergent and bleach. They consume an average of 30 gallons of water (20 gallons of hot water) at the maximum setting and 0.18 kWh of electricity to do a complete cycle.

Maytag has made a contribution to energy savings by offering a cold rinse option on all of its washers. Maytag gas and electric dryers are also energy savers, featuring a new drying principle that drys clothes faster: the Stream-of-Heat pattern. Warmed air travels diagonally from the upper right-hand rear of the drum to the lower left front, providing the longest possible flow of drying air within the drum. These Maytag units can dry loads 10 to 15 percent faster while using no more energy than do competitive dryers. The gas models from Maytag feature electric ignition systems rather than standing pilots to provide still greater energy savings.

Washers from the Speed Queen Division of McGraw-Edison have been designed to lower water usage, and lower wa-

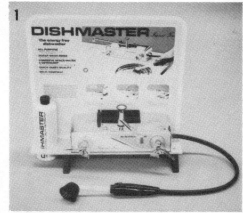

(1) Manville Dishmaster. (2) Presto WeeBakerie. (3) Whirlpool Model EET171HK Refrigerator.

ter consumption can mean reduced energy consumption— especially for users of the hot wash and warm wash cycles. Speed Queen dryers feature a vacuum drying principle to speed the drying process, thereby saving heating energy and costs.

The In-O-Vent is a device that attaches to your clothes dryer and retrieves heat and humidity normally vented outside. In winter, a clothes dryer not only vents heat used in the drying process, but it also draws heat from the room and dispels it outside as well. Therefore, the In-O-Vent can cut your winter heating bills by putting heat and moisture back into the home rather than venting it.

Ovens And Ranges

Pilotless ignition is one major way to save energy with a gas

range. The Magic Chef Model 346W-4HKPX features a glow-type ignition system instead of standing pilots, saving up to 30 percent of the energy required by other gas ranges. It also features a self-cleaning oven.

A Modern Maid QDU-796 pyrolytic self-cleaning range/oven can save up to one-third of the energy that a conventional oven consumes. This particular Modern Maid model also features a removable electric barbecue top.

Presto, makers of many small electric cooking appliances, offers a new one for the home bread baker. Called the WeeBakerie, it's just big enough for one loaf of bread. It produces the perfect baking temperature automatically; there are no controls to set. For someone who bakes bread one loaf at a time, the WeeBakerie certainly would provide energy savings over using a full-sized oven. (2)

Hot Water Heaters

THE WATER heater consumes a great deal of energy. In fact, in most homes it uses more energy than any cooking or refrigeration appliance, ranking second in energy consumption only to the heating system. Since the water heater is a sturdy appliance and since it's generally hidden away in the basement, however, many people tend to forget about it. But because it accounts for such a large percentage of your total energy use, your water heater deserves a good deal of care.

Check to see how well your water heater is insulated. A water heater tank that is not well insulated is an energy waster. Place your hand on the side of the tank. If the tank is warm, you know that heat is being lost that should remain inside. By wrapping insulation around such a water heater, you can start saving energy dollars immediately.

When applying the insulation (there are special kits made for

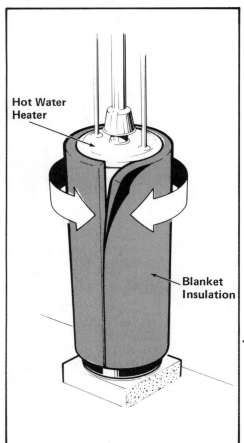

Hot Water Heater

Blanket Insulation

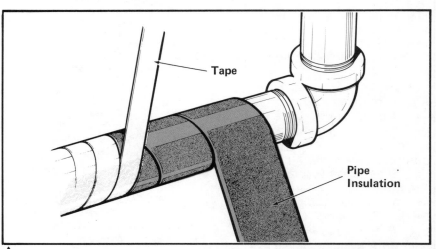

Tape

Pipe Insulation

▲
Wrapping hot water pipes with insulation prevents the water from cooling before it reaches the various fixtures where it is needed.

◄*You can start saving energy dollars by wrapping insulation around a water heater that loses its heat.*

There is no reason to have a water ► heater set to the hottest point on the control knob. Almost always the middle setting will suffice.

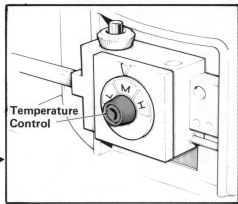

Temperature Control

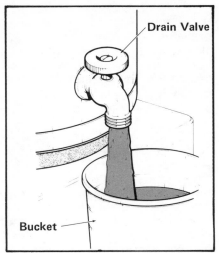

Drain the accumulated sediment out of water heater regularly to keep it functioning at peak energy efficiency.

this purpose), be careful not to cover the vents at the top of the heater. In addition, you may find that the tape supplied in the kits to bind the insulation around the tank fails to last long after being subjected to heat; you might wish to use bands of wire instead to secure the insulation.

If the pipes carrying water from the heater to the various fixtures are not insulated, a great deal of heat can be lost along the way. This is a particularly bad energy waster if the pipes travel through an unheated crawl space or basement. And in the summer, the wasted heat is doubly ex-

pensive because it makes your air conditioner work harder than it should. Easy-to-apply pipe insulation is available and should be installed wherever the pipes are accessible.

Many times, the thermostat on the water heater is set higher than needed. Bathing, showering, clothes washing, dish washing, and all the other household hot water uses don't require that water come out scalding hot from the tap. In fact, more and more fabrics can be washed in cold water, with many detergents formulated to clean laundry just as well in cold water as in hot. In other words, there is no reason to have your water heater set to the hottest point on the control knob. Almost always the middle setting will suffice.

If you plan to be away for a few days, turn your water heater off. And if you have an electric water heater, you can install a timer to prevent the unit from heating during the night or while there is no one home during the day.

Of course, as with almost every appliance, proper maintenance can help your water heater operate at its top level of efficiency. The most important maintenance involves draining the sediment that collects at the bottom of the tank. Sediment can act as an insulator, preventing the efficient transfer of heat from the gas- or oil-fired burner. In addition, sedi-

ment can cause the water heater to emit a rumbling noise.

Here's how to drain a water heater.

Step 1. Make sure that the drain valve turns easily. If it does not, it may have frozen from not being used for a long period of time. If it is frozen, attach a garden hose to the faucet so that when you do get the drain open and then cannot close it readily, you will be able to direct the water either into a drain or outside the house.

Step 2. Drain the tank early in the morning before anyone has used the hot water to insure that the sediment has settled to the bottom of the tank.

Step 3. Open the drain and let a few pints of water flow into a bucket or small container.

Step 4. Keep draining until the water runs clear. When it does run clear, your draining chore is done.

Once you do this task several months in a row, you will know how often it needs to be done. Sometimes, twice a year is all the draining required. By keeping the sediment out, you will have a more efficient and less noisy hot water system, and you will prevent future problems from developing in the tank and in the hot water pipes.

Hot Water Heaters

Materials and Supplies

Reduced burner rates, an improved baffle, more efficient insulation and a unique water inlet design are key reasons for the high energy efficiency of A.O. Smith's Conservationist gas water heater. The insulation is not

only thicker but also has improved thermal properties, while the water inlet design assures more uniform temperatures. A. O. Smith also markets a high-efficiency electric Conservationist water heater. (1)

Energy Miser water heaters from the Rheem Water Heater Division now employ a high-efficiency heater element called the Lifeguard. The outer sheath of the element is a tin-coated stainless steel material, while the interior packing consists of high-grade magnesium oxide protecting a high-density 80/20 nichrome heating filament. The element's unique low watt-density (rated at 70 watts per square inch) promotes longer life and more efficient operation. (2)

The new Gainsborough 7 Shower Heater is an easy-to-install water heating unit that saves energy. In contrast to a conventional hot water heater that operates continuously, this Gainsborough unit heats water only when someone is taking a shower, thereby eliminating standby heat losses. Neon indicator lights inform you of its setting, and a preset restrictor regulates the water flow. The Shower Heater connects directly to a ½ inch cold water line and needs just a 110 volt, three wire electrical supply to provide continuous hot water showering.

An instantaneous, at-the-tap water heater from Chronomite Labs can provide continuous hot water at any temperature between ambient and 212 degrees Fahrenheit. The small, electronic, boilerless water heater passes water between two electrodes, conducting current through the fluid by means of the electric conductive molecules in it. In effect, therefore, the electrical resistance of water functions as the heating element. (3)

The In-Sink-Erator's Steaming H_2O Tap, installed at the sink, provides 190-degree water instantaneously for making coffee, tea, soups, hot cereal and other fast foods. Installation can be a do-it-yourself project for a reasonably competent home handyman.

If you use a heat pump to heat and cool your home, you may be able to save on water heating. Friedrich has developed a small device that attaches to nearly any heat pump and that utilizes the discharged heat from the heat pump's compressor to heat water. A small pump in the unit circulates water from your existing water heater to the heat source.

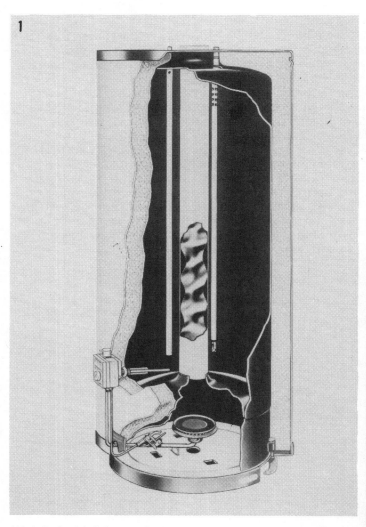
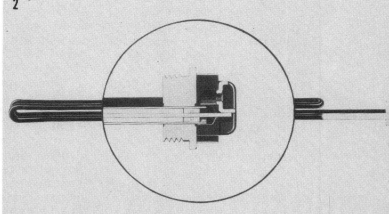

(1) A.O. Smith Conservationist gas water heater. (2) Rheem Lifeguard heater element for Energy Miser water heater. (3) Chronomite boilerless water heater.

The thermostatically controlled unit can supply about a third of your hot water needs. (6)

If you plan to replace your water heater, you'll be happy to know that you can buy flexible copper water heater connectors. EZ Plumb makes such connectors in various lengths from 12 to 24 inches. They eliminate sweat soldering, threading, require no special tools, and can be bent up to 180 degrees. These connectors are also good if you need to replace a broken connector pipe on your existing heater. (5)

Tank And Pipe Insulation

A water heater can lose a great deal of the heat it produces if the storage tank is not well insulated. If the tank feels warm to the touch, it's losing heat to the air surrounding the unit. Johns-Manville's Water Insulation Kit can put a stop to such energy wasting. Consisting of a blanket of vinyl-faced fiberglass insulation that fits around the heater tank, it keeps the heat inside the tank where it belongs. (4)

Any time you pay to heat water, it's a shame to lose much of the heat to a cold basement or crawl space before the water ever reaches the tap or appliance where it is needed. Foamedge insulating pipe wrap sleeves, from Teledyne Mono-Thane, are made of a polyurethane foam that conserves heat, stops sweating and reduces the sound level emanating from noisy pipes. The sleeves are slit along their lengths to make installation easy.

Wrap-On's Insul-Foil combines heavy aluminum foil with top-grade PVC vinyl foam to create an insulated wrapping for hot water pipes that prevents heat loss. The self-adhesive material can be applied to plastic pipe as well as metal, and it is easy to attach even around elbows and bends. Wrap-On also makes Insul-Strip—a version of Insul-Foil that comes in three-foot sections.

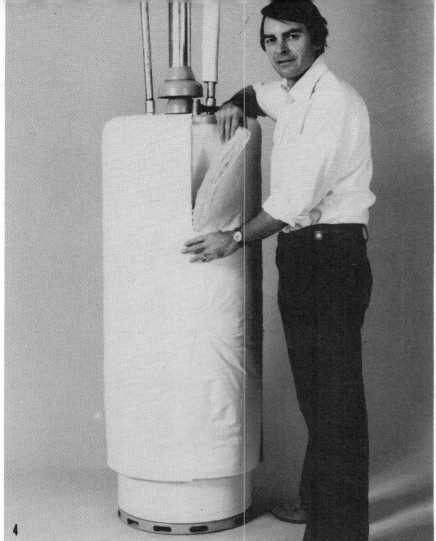

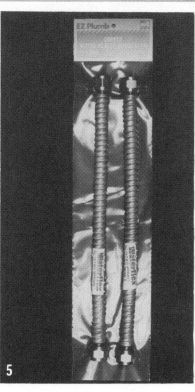

(4) Johns-Manville Water Insulation Kit. (5) EZ Plumb Water Heater Connectors. (6) Friedrich water heater device attached to heat pump.

Water Conservation

EVEN IF you live in an area where water is abundant, you should get into the water-saving habit. Lush green areas have been known to suffer severe droughts, but even if no drought occurs, water should not be wasted. It takes energy to treat and pump water, and since about a fourth of the water consumed in homes is hot water, cutting down on overall water use is certain to cut your energy costs.

The average family of four consumes more than 250 gallons of water every day for indoor use. The toilet—the biggest single user—accounts for about 40 percent. The bathtub/shower is next with about 33 percent, while laundry appliances take about 12 percent. Ten percent is used for cleaning, three percent for the lavatory, and two percent for drinking water and cooking. Of course, this is a general picture, and water consumption varies widely according to the lifestyle of the individual family. Keep in mind that this breakdown of water consumption covers indoor use only; it doesn't include outside water used for lawn sprinkling, car washing, etc.

There are many ways to cut down on the amount of water your family consumes. A host of new water-saving appliances, fixtures, and devices can do wonders, but water-saving consciousness on the part of individual family members is the most important single factor.

Since the toilet is the biggest user, it might be well to start there. Be sensible—don't flush every time you toss in a dead bug, a cigarette butt, or tissue. On the other hand, avoid some water-saving techniques that do more harm than good. Putting a brick in the toilet tank, for example, is often more harmful than beneficial. Some bricks start to disintegrate through prolonged contact with water, leading to potential damage to the commode that could waste water and certainly cost you a great deal of money to repair. A plastic jug filled with sand would displace more water than the brick, but remember that any time you place an object in the tank, you must make certain that it does not interfere with the operation of the toilet itself. Otherwise, it can cause you to waste more water than you save.

If you are determined to reduce the amount of water that your toilet consumes, investigate some of the add-on water-saving

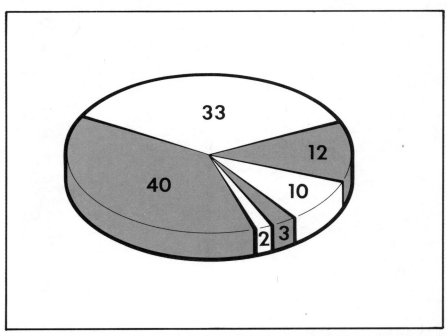

Of the 250 gallons of water consumed daily by the average family of four, the toilet accounts for about 40 percent, bathtub/shower 33 percent, laundry 12 percent, cleaning 10 percent, lavatory 3 percent, and drinking and cooking 2 percent.

devices now on the market. And if you are building or remodeling a bathroom, consider the new water-saving toilets designed to use less water when flushed.

Bath water can be conserved by installing low-flow shower heads and by turning the water off while soaping.

Both laundry and kitchen water savings depend upon the proper use of appliances. Wait until you accumulate a full load before running either the dishwasher or clothes washer. If you hand wash, keep the water off except when you absolutely must have it on.

You can replace all faucets with low-flow attachments, although prudent use makes the most sense. Of course, leaking faucets and other water-wasting situations should be remedied immediately.

Fixing A Dripping Faucet

AMONG THE MORE harmless of household problems is the dripping faucet. Most people reason that a drip just amounts to a few drops of wasted water, so why worry? Yet, if you stop to add up how much water is wasted in a year, the cost can mount up to around $50 right down the drain. If you call for a plumber to fix it, of course, he will charge you $15 to $25, but you can repair a dripping faucet in a few minutes, and the parts cost no more than a dollar. Follow these steps.

Step 1. Shut off the water at the cutoff below the sink or at the main cutoff where the water supply pipe enters your home. Open a faucet at the lowest point in the house to drain water out of pipes.

Step 2. Remove the packing nut with an adjustable wrench. You may first have to flip up a button,

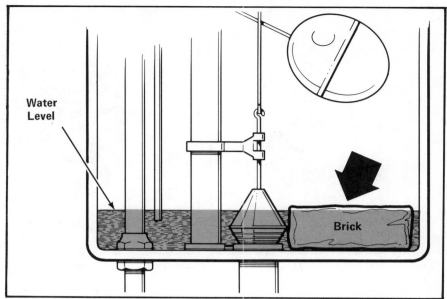

Putting a brick in the toilet tank is often more harmful than beneficial. The brick will not save much water, and it could damage the toilet if it starts to disintegrate after prolonged submersion in the water.

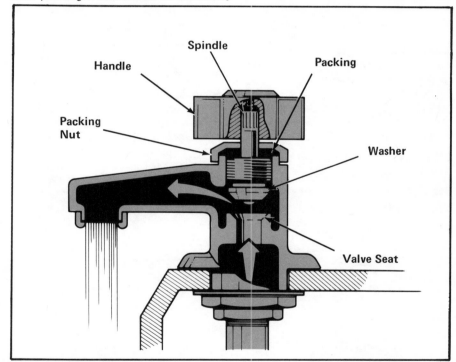

▲
Most faucet leaks are caused by poor contact between the rubber washer at the base of the spindle and the valve seat. Replace the washer or repair or replace the valve seat.

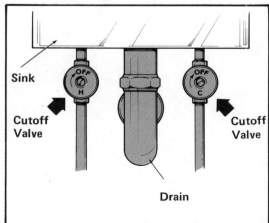

◄ *Before attempting to fix a dripping faucet, shut off the water at the cutoff below the sink or at the main cutoff where the water supply pipe enters your home.*

123

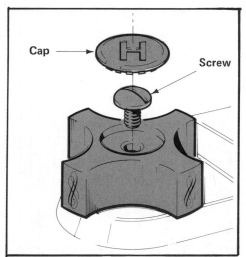

To reach the screw holding the faucet handle, you may have to flip up a button.

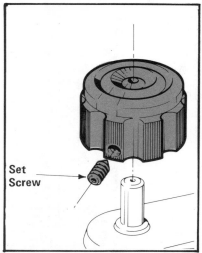

Some faucet handles are held by set screws which must be removed.

handle to screw the tool down against the seat, causing the cutting teeth to grind the seat smooth. Make sure that the seat is smooth and shiny after using the tool.

Step 7. Coat the threads of the spindle with petroleum jelly.

Step 8. Reassemble the faucet, and turn the water back on.

If your faucet leaks around the handle only when the water is turned on, you need to replace the packing. Here is how to do that simple repair procedure.

Step 1. Remove the spindle.

Step 2. If you have not already removed the handle to get to the packing nut, remove the handle now.

Step 3. Slide the packing nut up off the spindle.

Step 4. The blob under the packing nut is the packing. It may be a solid piece, or it may be a string of black graphite material that is self forming. Replace the old packing with new packing of the same type.

Step 5. Coat the threads of the spindle with petroleum jelly.

Step 6. Reassemble the faucet, and turn the water back on.

remove the screw under it, and slip the handle off to expose the packing nut. In some cases, you will find a set screw holding the handle; remove it and the handle to get at the packing nut. If the packing nut is highly visible or made of chrome, protect it from the wrench by wrapping it with masking tape.

Step 3. With the packing nut off, turn the spindle out.

Step 4. At the bottom of the spindle you will see the washer held in place by a brass screw.

Step 5. Remove the brass screw, and replace the washer with a new one of the same size. Reinstall the brass screw.

Step 6. Inspect the seat down in the faucet. If it is scarred or corroded, either clean and reface it with an inexpensive reseating tool or replace the seat itself. The reseating tool or valve seat grinder has cutting teeth. Insert the tool in the faucet and install the packing nut over it; then turn the

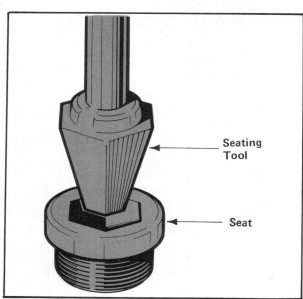

You can remove a faulty valve seat and install a replacement with an inexpensive valve seating tool. The seat is threaded for easy replacement.

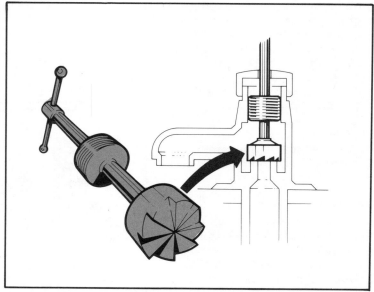

A valve reseating tool is designed to clean and reface a scarred or corroded valve seat. Its cutting teeth can grind the seat smooth and shiny.

Hints And Tips On Better Water Usage

MANY PEOPLE have taken water for granted for so long that they never think in terms of saving it. Here are some common sense rules for reducing water consumption in your house.

1. Since the toilet is a big water user in most homes, you should consider adding one of the several toilet water-saving devices presently available. Some permit adjustable flushing, allowing for a partial flush for non-solid waste and a full flush for the solids.

2. Never flush a toilet just to get rid of a cleansing tissue or a cigarette butt.

3. To cut down on water consumption at the lavatory when brushing your teeth or shaving, turn the water off and on as needed; don't just let it run.

4. Keep a pitcher of drinking water in your refrigerator instead of turning on the kitchen tap and letting the water run until it gets cool.

5. Run your dishwasher and clothes washer only when they contain full loads.

6. Use a bucket of water and a sponge to wash the car instead of turning on the hose.

7. Use a broom to sweep walks and drives instead of hosing them down. If wet cleaning is necessary, use a broom or scrub brush dipped frequently in a bucket of water.

8. When watering the lawn, make sure that your sprinkler does not waste water on walks and other unplanted areas.

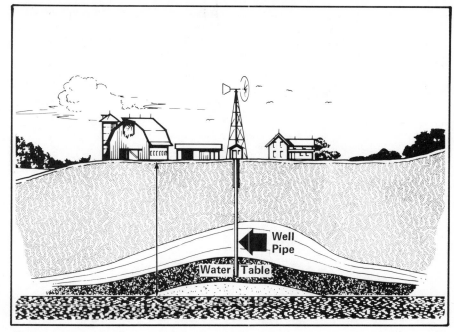

Frequently, a ground water reservoir (known as the water table) is not very far below the surface, and by tapping into the top of a soil or rock strata many people are able to pump enough water for their own use.

Drilling Your Own Water Well

INSTEAD OF depending on a utility company or community water works for your water supply, consider drilling your own water well. Many rural homes get their water from wells, and there are entire communities that depend on well water to meet their needs. In some areas ground water is not very far under the surface, and by tapping into the top of a soil or rock strata which has ground water, you may be able to pump up enough water for your use. The top of a ground water reservoir is known as the water table.

Before you run out to start digging, however, you should know what you're doing. There are many areas where ground water is either too deep to recover or just not there. In addition, not all the underground water is safe for drinking, and there are also many state and local regulations governing the drilling of wells.

The first thing to check before you start drilling a well is the level of ground water. Unless you are a geologist, consult the county agricultural department or a similar office. If you live where water wells are known to be practical, there should be several drilling contractors nearby who can tell you the general depth of the water table and explain the regulations (if any) regarding your proposed well.

Should you find water and drill a well, the likelihood of the water being safe for drinking is good. The soil that the ground water flows through acts as a filter, providing natural purification. On the other hand, water that flows in pockets of rock isn't filtered, and pure water can be made unsafe if it collects insecticides or if sewage systems drain into it. Of course, even if the well water is not acceptable for drinking, it can still be used for irrigation of crops or watering of lawns.

Years ago, wells were generally of the dug well type—i.e., a hole or pit, usually at least three feet in diameter, dug by hand.

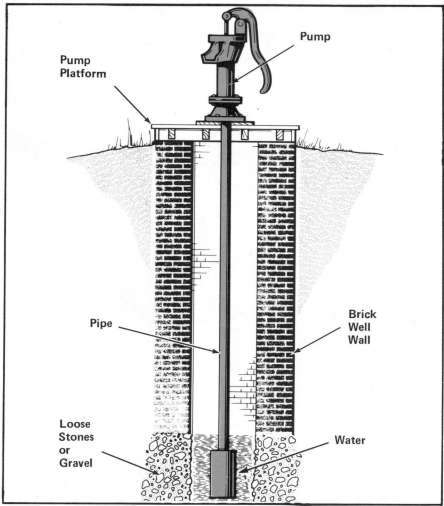

Old-fashioned wells were often dug by hand and then lined with bricks, rocks, wood, and other supporting materials. The area below the water table was generally lined with loose stones and gravel through which the water seeped.

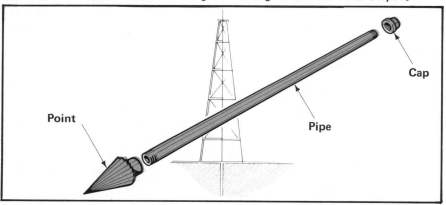

The newer driving method for drilling a well requires a device called a well point that is attached to a pipe; the pipe later becomes the well casing.

The depth was whatever it took to reach the water table. Because hand digging was such a hard task, these wells generally were not dug where the water table was very deep. The hand digging process was also quite dangerous due to the possibility of the sides caving in. Once the well was dug, therefore, the sides were lined with bricks, rock, wood, and other supporting materials. The portion of the well below the water table was generally lined with loose stones and gravel which allowed the water to seep in.

The newer methods for drilling a well—the boring method and the driving method—are much safer and easier.

Boring can be done with a hand auger (much like a posthole digger) to which extensions are added as the depth of the well increases; the handle of the tool is merely moved to the top of the extension. The auger is turned into the ground until its bit is full. It is then lifted out and emptied before being put back in the ground for more turning. Naturally, the boring method is suitable only for shallow wells, and it is best used in clay where cave-ins are not a problem. There are special auger spirals that replace the regular auger for rock removal, but if rocks are a big factor, you should choose another method of well digging.

Driving requires a device called a well point that is attached to a pipe. Sections of pipe are added as the depth increases, and a drive cap is placed over the ends of the pipe to protect the top end. A hand driver is usually lifted by pulley and wench. When the digging is done, the same pipe that was used in driving then functions as casing for the well.

Either method, of course, demands a great deal of hard work. In some areas, you might be able to rent a rotary drilling rig which, naturally, makes the process much easier. On the other hand, hiring a professional driller to dig the well usually costs little more than the rental of a rotary drilling rig, and having the job done professionally should be considered as a practical alternative to digging the well yourself.

If you are going to hire someone to do the drilling for you, though, be sure to check with the Better Business Bureau and other local consumer agencies. Ask the driller for recent references and check them out. In addition, before you select the person to do the drilling, get several estimates, and then make sure that the contract spells out everything that is to be done.

Water Conservation

Materials and Supplies

Water-Saving Faucet Attachments

The Conservarator from Baron Industries is a replacement aerator designed to slow down the flow of, and thus save, water. Even if the water pressure in the line varies, the Conservarator keeps water coming out in a constant flow. The unit comes in several different models to fit every size faucet in both kitchen and bath, and installation is a quick and easy task. (1)

Wrightway's Bubble Stream water saver kit includes an aerated shower head, an aerator for the kitchen sink and another for the lavatory. All three pieces in the kit are easy to install as replacements for existing fittings. These low-volume aerators can result in a significant reduction in water consumption. (3)

The Flo-Down controls built into all of the new Price Pfister faucets and shower heads keep water flow below three gallons per minute at normal water pressures. Older faucets can be retrofitted with adapters to accommodate the control units, which will also fit many faucets and shower heads made by other manufacturers.

The Nova replacement shower head, a product of Ecological Water Products, Inc., can reduce the water flow by as much as 66 percent. Even with this reduction however, the water still comes out in a forceful spray. Nova also markets a water saver aerator for kitchens and lavatories that cuts the flow by about the same amount without loss of an effective spray. (2)

(1) Conservarator replacement aerator and flow restrictor. (2) Nova replacement shower head. (3) Bubble Stream water saver kit.

American Standard's Colony Shower Head cuts water usage from approximately 3½ gallons per minute to only 2½ while allowing the user to adjust the spray from fine to coarse without sacrificing comfort. The chrome swivel head and plastic nozzle are easy to remove for cleaning. (4)

T&S Brass and Bronze Works markets a foot pedal control for use with kitchen faucets that lets you turn on the flow only when needed. This device can save a great deal of water for people who wash dishes by hand. (6)

Toilet
Water Savers

Flip the handle up on an Eco-Flush unit for a partial liquid-waste flush and down for a full flush to remove solid waste. Easily installed in most toilet tanks—replacing the present trip lever, handle and the entire tank ball assembly—the Eco-Flush is constructed almost entirely of plastic and will not rust or corrode. Since 80 percent of all flushes are for liquid waste and since Eco-Flush saves about three to four gallons during a partial flush, this device can save a substantial amount of water.

Moby Dike from Watersavers Inc. and Little John made by Metropolitan Watersaving Company Inc. employ a pair of flexible plastic plates with gaskets to create miniature dams inside the toilet tank. When the toilet is flushed, the water outside the two plates remains inside the tank. Best of all, no appreciable reduction in flushing efficiency occurs in most toilet tanks. Moby Dike, Little John and other water saver dams currently available are not designed for use in the new smaller toilet tanks, the si-

lent flush toilets, or any tankless system. (9)

Most toilets use from five to seven gallons of water per flush, but the Geberit tank uses only 2½ to 3½ gallons, depending on the model. In addition to using only a small amount of water, the Geberit tank possesses other unique features. Both the float unit and the flush valve unit are hydro-balanced for faster operation in flushing and filling; operation is extremely quiet, and the plastic tank is styro-insulined to prevent condensation. Geberit tanks are available in models to fit just about all of the best-selling toilets in the United States. (8)

The Microphor toilet uses only about two quarts of water per flush. Compatible with most existing home plumbing (it has Uniform Plumbing Code approval), this toilet combines air pressure with water and waste in an

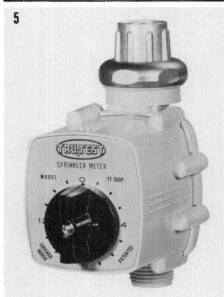

evacuation hopper to transmit waste to a regular sewer line. (10)

Due to the water shortage, Flushmate designed a new tank to cut water usage. Water entering the tank compresses the air inside, and when air and water pressure are equal, the water flow stops. When the flush button on the Flushmate is pushed, the main valve inside the tank is lifted and the compressed air forces the water through the bowl. Complete flushing requires only about two gallons of water. Although the Flushmate tank is about five inches lower than conventional tanks, replacement is a do-it-yourself project.

Borg-Warner's Artesian Water-Saver is a slightly altered toilet design that uses only 3½ gallons of water per flush, compared to about five gallons per flush with conventional siphon-jet toilets and as much as 6½ gallons per flush with other types. The Water-Saver operates just as efficiently as the conventional toilets, however.

The installation of a Won-Dare Water Saver from Dare Pafco, Inc. converts the toilet for a partial flush when only liquid waste is in the bowl, while allowing a complete flush when solid waste is involved. The all-plastic unit can be installed fairly easily, although some adjustment must be made for different types of toilet tank mechanisms.

Lawn Sprinkler Meters

If you're the forgetful type who lets the sprinkler remain on too long, a Tru-Test Sprinkler Meter can save you a great deal of water. You just set the amount of water needed, and the meter shuts the sprinkler off automatically when this limit is reached. The Sprinkler Meter can be set for any amount up to 1500 gallons. (5)

Water Well Drilling

A low-cost one-man drilling rig called the Hydra-Drill is being marketed by the Deeprock Manufacturing Company. Capable of drilling down to 200 feet, the rig is capable of speeds up to 40 feet per hour. Water from a garden hose attached to the drill flows down through the mechanism stem to wash up cuttings as you drill. When and if you strike water, you remove the drill bit and replace it with a reamer bit to widen the hole. Next you insert the casing and attach a pump; now you can start using your well. (7)

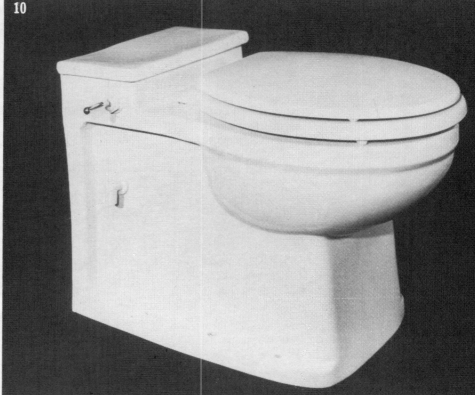

(4) Colony Shower Head. (5) Tru-Test Sprinkler Meter. (6) T&S foot pedal control. (7) Hydra-Drill. (8) Geberit tank. (9) Little John. (10) Microphor toilet.

Automobile Energy Savings

IN THE AVERAGE American family, the automobile is responsible for more energy consumption than any other single entity. Since this statistic takes into consideration the many urban residents who use public transportation, those who drive cars frequently are spending a truly large percentage of their total energy expenditure on their automobiles.

Without knowing anything about the mechanical aspects of how a car functions, you can save a significant amount of fuel every time you drive. And since every ounce of gasoline you conserve is money in your pocket, you should heed the following suggestions.

1. Start smoothly from a stop; blast-offs waste gas. On the other hand, get to driving speed as quickly as possible because lower gears use more fuel.

2. Drive at a moderate speed. Most cars and trucks get better mileage at 55 mph than at 70 mph.

3. Avoid stop-and-go acceleration. A bobbing speedometer indicates fuel waste.

4. Don't sit with the motor running for more than a minute. Since it will take less gas to restart the engine than to let it idle for any moderately lengthy period, turn it off.

5. Avoid overfilling the gas tank. Expansion and simple sloshing can lead to costly spillage.

6. Keep your tires properly inflated. A low tire makes the engine work harder.

7. If you must stop while going up a steep hill, don't use your accelerator to hold the car still; that's something your brakes can do very well.

8. Use your car's air conditioning system as little as possible. Fuel consumption goes up by ten percent or more when the air conditioner is on.

9. Avoid prolonged winter warm-ups.

10. Keep your left foot off the brake pedal while driving; dragging brakes make your engine work much harder.

11. Be sure your chassis and wheels are properly aligned.

12. Have your transmission checked. An automatic transmission that slips can cause your car to waste lots of gasoline.

13. Keep your car's engine properly tuned.

14. Car pool wherever possible, plan necessary car trips carefully so as to accomplish all errands with a minimum of driving, and try walking rather than driving to all nearby locations.

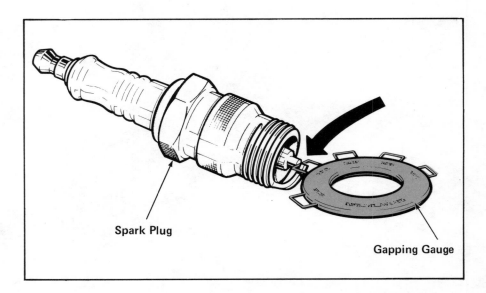

Spark Plug

Gapping Gauge

Spark plugs, even new ones, must be properly gapped. A gapping gauge makes the task quite simple. Your car owner's manual has the gapping specifications.

Tune-ups

A PROPERLY tuned engine operates at peak efficiency, saving as much as ten percent of the fuel that a car in need of service requires.

The term "tune-up" means different things to different people. It usually involves changing the spark plugs and points, but it can include replacing many other parts that play a role in a car's energy efficiency. The best guide to your car's tune-up needs is your owner's manual; if you don't have one, contact your authorized dealer. The manual will tell you how frequently (in either time or mileage) to service your particular vehicle. The following references to service intervals should, therefore, be considered average or typical; they may not be appropriate to your make and model.

Spark Plugs. Change spark plugs every 10,000 to 15,000 miles, installing quality plugs of the proper type and size. Auto supply departments have charts telling you what plugs your car requires.

Spark plugs, even new ones, must be properly gapped; the owner's manual contains the appropriate information. If all your driving is at low speeds, you should clean your plugs and check the gap about half way between changes. Changing plugs is an easy do-it-yourself project that most car owners can accomplish. The only special tools needed are a spark plug wrench (or spark plug attachment for a standard socket wrench) and a gapping gauge.

Ignition Points. Cars that are not equipped with electronic ignitions generally need the points changed at the same time the spark plugs are replaced. Chang-

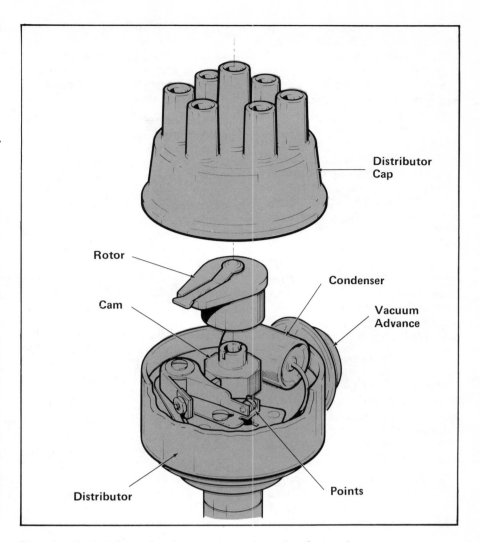

Changing the ignition points is very easy and requires few tools.

ing of points is very easy, requiring only a screwdriver and feeler gauge, but the timing and dwell—which must be set whenever the points are changed—necessitate a timing light and dwell meter. Unless you have access to these instruments, you'll need to take your car to a shop for service.

Condenser. Most points come in a kit with a condenser included, although the condenser usually does not need changing as often as the points do. Since the cost factor is minimal and the replacement procedure is easy, however, smart owners change the condenser when putting in new points.

Air Filter. A clogged air filter can seriously reduce a car's gas mileage. Check the filter by holding it up to a strong light. Gener-

ally, you can see if it is clogged with dirt. In areas with fine sand or dust, however, the clogging may not be visible. If you live or

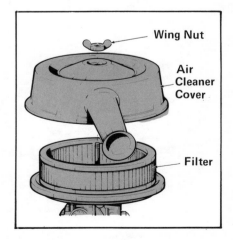

Check and change the air filter at regular intervals. Replacement is simple.

drive in such dusty areas, therefore, change the filter twice as often as your owner's manual suggests.

Although most service stations will not charge you labor for putting in a new air filter, they charge much more for the filter itself than what you can buy one for at many auto supply stores. Therefore you can save money by simply installing the replacement air filter yourself, and it is a simple job. The auto supply store has a chart telling you the right size filter for your car.

Timing. A timing check is a must when replacing or adjusting points. In addition, if your engine lacks power, runs rough, or knocks and pings, you might suspect the timing.

Vacuum Advance. This is a task for a competent mechanic. Done at the same time the timing is set, it usually requires only an adjustment, but a car with more than 30,000 miles may require replacement of a diaphragm assembly.

Cylinder Head Tightening. A loose head gasket can reduce engine power, and although the only task involved is tightening a bolt, it is a job for a mechanic. The bolt must be tightened with a torque wrench to the precise specifications set forth by the car manufacturer.

PCV Valve. Replace this emission control device every 12,000 miles, and clean the hoses to the valve at the same time. When the system gets clogged, it can cause stalling and engine roughness in addition to wasting fuel. Remove the valve from the engine cover, and with the engine idling, check for suction by placing a finger over the openings in the valve. No suction or very weak suction means a clogged system.

Spark Plug Wires. Normally, spark plug wires last for 24,000 miles, and although they might endure beyond this point, they will be on the road to deterioration. Check the wires with an ohmmeter when changing the plugs or at the first sign of backfiring. Be sure to buy quality replacement wires of the proper type and to remove and replace each wire before removing another.

Carburetor And Fuel System

THE FUEL system, of course, is where significant energy saving—or wasting—can occur. Look for leaks. Gas tanks, lines, and other parts of the fuel system are subject to leaks, and it is just common sense to have these leaks repaired immediately. In addition, a missing or faulty gas cap can create a safety hazard as well as result in lost gasoline; get a replacement cap without delay.

The most crucial component in the fuel system insofar as gasoline saving is concerned is clearly the carburetor. The carburetor is a complicated assembly, and even though carburetor overhaul kits featuring step-by-step instructions are available, the average motorist should take his or her car to a competent mechanic when a carburetor overhaul is indicated. On the other hand, there are several things that almost anyone can do to keep a carburetor from malfunctioning.

Cleanliness comes first. A clogged air filter, for example, can allow dirt to enter the carburetor. In addition, gummy residue deposits from gasoline can cause carburetor problems. Wherever the dirt comes from, once it reaches the carburetor, it can cause hard starting, flooding, poor acceleration, stalling, fouled plugs, rough running, and poor gas mileage.

You can clean your carburetor by pouring a specially formulated chemical into the gasoline tank; the chemical then works its way into all parts of the carburetor and fuel system as you drive the car. Other cleaners can be introduced directly into the carburetor while the engine is running; it is only necessary to remove the air cleaner and spray the cleaner all over and inside the carburetor. Unlike the gas tank solutions, however, the

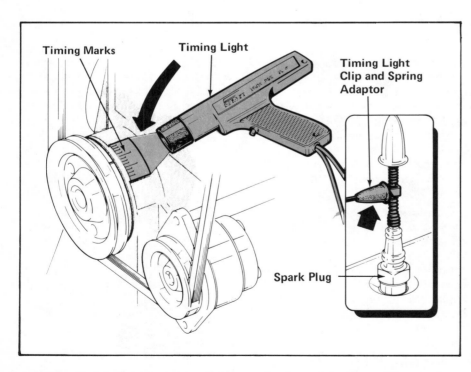

Timing Marks

Timing Light

Timing Light Clip and Spring Adaptor

Spark Plug

A timing check is a must when replacing or adjusting points. You will need a timing light and a dwell meter, however, and unless you have access to these tools, you'll have to take your car to a shop for service.

spray cleaners remove grime only from the parts of the carburetor that are accessible to you with the aerosol spray can.

Dirt and gummy deposits can also foul the automatic choke, which plays an important role in cold weather starts. A cold engine needs a richer fuel mixture, which means more gas and less air. The automatic choke closes a butterfly valve in the carburetor to control this mixture. If the valve sticks, the car may not start at all or at best be very difficult to start. Vigorous pumping of the accelerator pedal can often help, but such pumping frequently floods the carburetor, wasting gasoline. You can clean the automatic choke with an aerosol spray cleaner made for this purpose. Spray the external linkage too.

If your car has an in-line filter or one that is part of the fuel pump, have the filter replaced about every 12,000 miles. Other gas filters, located in a recess in the carburetor just where the fuel line comes in, should be replaced about every 6000 miles.

The fuel pump is sometimes a source of leaking fuel. If you detect any sign of gasoline leaking, have the pump and all fuel lines checked immediately.

Gas-Saving Gadgets

THE FUEL shortage has proved a boon to manufacturers of "miracle accessories" for your car. For example, there are replacements for conventional spark plugs that claim to increase your mileage, and there are special gas-saving ignition systems, gas-consumption gauges, speed- and throttle-control systems, and air resistance spoilers—all designed to make your car run farther on a gallon of gas. Naturally, some are worthwhile and others (many others) are not.

One true gas saver you can buy for your car is a set of radial tires. Tests prove that radials offer less resistance than other types of tires and, therefore, increase your mileage per gallon. Although radials generally cost more than other types of tires, the price differential will prob-

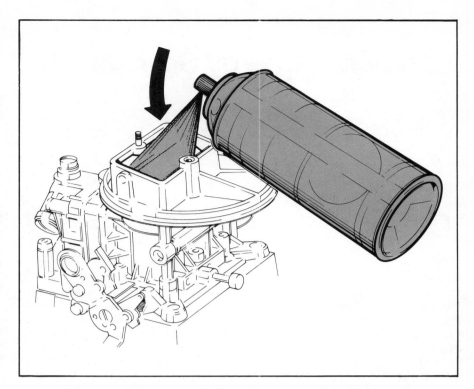

Gummy gasoline residue can foul the carburetor and result in poor gas mileage. Spray an aerosol carburetor cleaner to remove a great deal of this dirt and grime.

ably be offset by the fuel savings over the life of the radial tires. As is the case with any tires, you must keep radials properly inflated for long tire life and maximum fuel efficiency.

An electronic ignition system can offer some advantages, but a non-electronic ignition system in tip-top shape delivers nearly as good performance. From the standpoint of increased gas mileage, therefore, installing an add-on electronic system would probably not be worthwhile.

The spark plug replacement devices have not proved to offer any advantage over properly gapped conventional plugs.

Throttle- and speed-control (or "cruise-control") devices do help to conserve fuel because they maintain your car's speed at a constant level, preventing fuel-wasting pumping of the accelerator. If you do a great deal of highway driving, you should certainly consider having one of these control devices added to your car.

An accurate gas-consumption gauge—a vacuum gauge calibrated to show fuel use—can improve your driving habits and let

you know when your engine isn't performing properly. The type that indicates only when gas is being wasted can help you drive so as to maximize fuel efficiency. There are gauges, however, with calibrated dials that are a great help in diagnosing gas-wasting engine ills. No matter which type of gauge you choose, be sure to get a lighted unit.

Spoilers or wind deflectors reduce drag and thus improve mileage, but the need for reduced wind drag varies so much with different vehicle designs that it is impossible to gauge gas savings accurately. Long-distance truckers who add spoilers generally experience significant mileage improvements. On a family car these add-ons may never pay for themselves.

Although the flood of gas-saving gadgets shows no sign of receding, most of the "miracle devices" presently available claim more than they actually deliver. In most cases, properly tuning a car is more worthwhile and presents a better return on your gas-saving investment than buying some "magic accessory" that claims to triple your gas mileage.

Automobile Energy Savings

Materials and Supplies

Ignition System

To maintain an efficient automobile, the home mechanic should check over the engine periodically. And, because some engine parts are controlled by vacuum, a vacuum gauge is a very useful instrument. It is about the only way you can tell where pinpoint leaks may be. The Model CP 7801 vacuum tester, made by the Sun Electric Corp., features an easy-to-read dial in a chrome-plated housing, a 48-inch neoprene hose, T-fitting, 1/8 and 3/8 inch manifold fittings and a universal adapter. The device also is able to measure your fuel pump pressure.

To provide your vehicle with greater efficiency and let you do away with points and condensers, you might consider installing one of the electronic conversion kits on the market. One made by Accel Eliminator Ignition can be installed without the need of special tools. The Accel conversion kits feature quick-connect wiring harnesses and color-coded wiring to prevent improper hookup. (1)

Poor ignition robs your automobile of efficiency. Filko Automotive Products Division of F & B Manufacturing Co. makes replacement parts such as trigger wheels and power modules for American Motors, Chrysler, Delco and Ford igniton systems.

You can save fuel and improve the efficiency of your car if you give your vehicle a regular tune-up. The Blue Streak heavy-duty tune-up kit, made by Standard Motor Products, Inc., contains everything you need— points, rotor and condenser—for all American-made cars.

An energy-saving tune-up is never complete until the engine

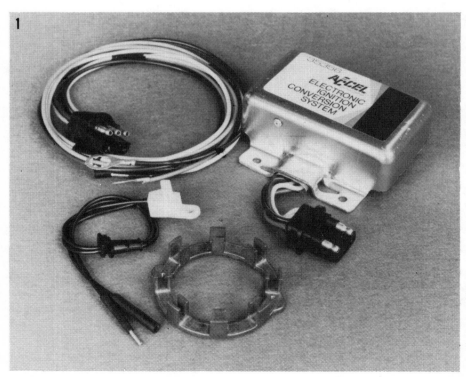

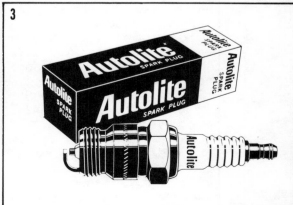

(1) Accel electronic ignition conversion kit. (2) Sun Electric Inductive Timing Light. (3) Autolite spark plug. (4) Automotion Silicone Insulated ignition wires. (5) Filko spark plug cable remover. (6) Siloo tune-up and PCV solvent.

timing is set properly. One of the easiest timing lights to use is the Model CP 7501 Inductive Timing Light made by the Sun Electric Corp. It is connected merely by clamping the unit to the spark plug wire. The instrument also features reverse polarity protection and a rubber nose cone. Sun also manufactures a combination dwell-tach and inductive timing light. (2)

The spark plug is the last step in the car's ignition circuit. If everything is working properly and the right fuel is being used, even burning of the compressed mixture results, the crankshaft, turns, and the engine is running efficiently with maximum mileage and lowest pollution emissions. But when the service life of your spark plugs is over, replacement is a must. Spark plugs to fit all vehicles are manufactured by the Fram Corporation (Autolite), the Champion Spark Plug Co., and the AC Spark Plug Division of General Motors Corporation.(3)

Due to the intense heat produced under the hood of modern cars, spark plug, or ignition, wires can split and wear out, resulting in backfiring and energy waste. So spark plug wires should be checked whenever you change plugs during a tune-up. Automotion Silicone Insulated ignition wires, manufactured by Resistance Products, Inc., feature spring-lock plug terminals and silicone spark plug boots. (4)

To remove the spark plug wires without damaging the spark plug boots, you need a spark plug cable remover. Filko Automotive Products Division of F & B Manufacturing Co. makes a spark plug cable remover, Model 7010, that makes this job easy. This tool is especially helpful when you are working on a hot engine or where the spark plugs are difficult to reach. (5)

Fuel System

For every gallon of gasoline your car burns, it must breathe thousands of gallons of air. If the air is not able to reach the engine due to a dirty air filter, the vehicle will be gasping for breath and efficiency will be reduced. Air filter replacement is easy and less expensive than having it done at your service station. Replacement air filters by the Fram Corporation are widely distributed and available for most cars on the road today. But whatever brand your dealer carries will be better than a dirty air filter in your car.

If your PCV valve is not too far gone, the easiest way to clean it is while you drive. Siloo, Inc. makes a tune-up and PCV valve solvent formulated to give your car an on-the-road tune-up. The product is designed to dissolve gum and sludge deposits and to free sticky valves. Such problems increase gas use. (6)

The purpose of a positive crankcase ventilation (PCV) valve is to ration unburned hydrocarbon fumes from the crankcase into the carburetor, where they are rerouted into the combustion

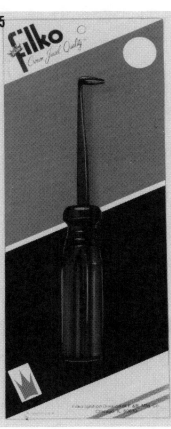

process. A dirty valve will make the engine run rough and consume much more gasoline. It can increase oil consumption too. Some firms that make PCV valves include the Wells Manufacturing Corp. (Ampco) and the Carol Cable Company Division of Avnet, Inc. Replacement is easy, but make certain you obtain the proper type for your make, model and year of automobile. (7) (9)

A dirty or clogged fuel filter can cause you headaches on the highway, but replacement is easy and inexpensive. Fram Corporation and Filter Dynamics International (The Lee Eliminators) both market fuel filters that are distributed nationally. Auto sup-ply stores carrying these brands will usually carry filters for virtually all makes and models of cars. Replacement instructions are on the packages.

A leaking fuel pump wastes gasoline and a malfunctioning pump can immobilize your car. When you need to replace a fuel pump, a rebuilt model can save you money. Ren-O-Vated fuel pumps, made by Kem Manufacturing Co., Inc., are available at auto supply stores. Most internal components are completely replaced, and after it is rebuilt, the pump is thoroughly tested.

The heart of your vehicle's fuel system is the carburetor and it should always be in top condition. Even though it is a complex part of your car, it is possible for home mechanics to overhaul this component with a Jiffy tune-up kit, a product of the Hygrade Products Division of Standard Motor Products, Inc. It contains the parts you will need and illustrated step-by-step instructions. Other kits with detailed instructions include Ampco Carburetor Tune-Up kits from Wells Manufacturing Co., Auto-Mech kits from Allparts, Inc., and Filko kits made by F & B Manufacturing Co. (8)

To make your vehicle perform more efficiently, its components should be clean. This is espe-

(7) Carol Cable replacement PCV valve. (8) Hygrade Jiffy tune-up kit. (9) Wells PCV valve. (10) Gumout carburetor cleaner. (11) Duro Special Agent No. 3. (12) Hi Temp gas line antifreeze. (13) Choke-Ease choke cleaner. (14) Stant locking gas tank cap. (15) Thextonite Gas Tank Sealer Stick.

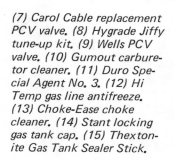

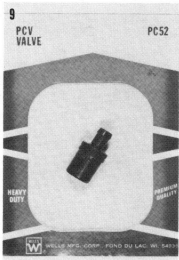

cially true for the carburetor and automatic choke. There are many good products on the market for this purpose. Aerosol carburetor cleaners are marketed by the Valvoline Oil Co., Woodhill/Permatex, The Maywood Co., Wynn's Friction Proofing Supply Co., Marvel Oil Co., Pennzoil Co. (Gumout), the Shaler Co. (Rislone), and Stewart-Warner Corp. (Alemite CD-2). Aerosol sprays also are available for cleaning linkage and automatic chokes. They include products made by the American Grease Stick Co., (Choke-Ease), Woodhill/Permatex (Duro Special Agent No. 3), and Stewart-Warner (Alemite CD-2). There also are gasoline

additives, such as STP, to help clean carburetors. (10) (11) (13)

Although your gasoline has protection against gas line freeze, extremely low temperatures can cause such freezing if moisture is present, reducing efficiency. Extra help is available in the form of gas line antifreezes such as Hi Temp from the Gold Eagle Co. It is safe to use with cars equipped with catalytic converters. (12)

If your gas tank is leaking, you are not only wasting fuel but you have a potentially dangerous situation on your hands. Thextonite Gas Tank Sealer Stick, made by Thexton Manufacturing Co., can

quickly repair minor leaks without draining the tank. The self-polymerizing sealant, applied from the outside, is merely rubbed over the leaking area. (15)

An energy crisis of another kind can be caused by a thief who siphons gasoline from your tank. A locking gas tank cap could prevent this. There are many such caps on the market, and Stant Manufacturing Co., Inc. has a line that will fit most cars and trucks. All caps are emission control type and meet or exceed original equipment specification. A locking cap also stops vandals from contaminating your tank. (14)

12

13

14

15

Alternate Sources Of Energy

RIGHT NOW, the greatest contribution most Americans can make to easing the energy crisis is by conserving existing fuel sources. If everyone took all the steps described in this book, the life of known fossil fuels could be extended by decades. On the other hand, since those fuels are certain to run out eventually, new energy sources must be tapped to replace the depleted ones. Without question some of the sources currently in the experimental stage will become crucial to your way of life before long.

Solar Energy

MANY PEOPLE feel that the sun will be the source for much of our future energy. While that prediction may come true, at present solar is limited to supplementing more traditional sources. Solar energy is in the testing stage now, with technological developments changing the picture on nearly a daily basis. As expected, not everything being tried works out, and while there are actual solar energy systems working in homes today, these systems are not yet mass produced and therefore are too costly for most Americans.

On the other hand, what is now expensive and "iffy" could become an economical and reliable way to save energy in the very near future. Be aware, but be wary.

The first thing you must realize is that nearly all home solar energy systems require a supplemental system of the conventional type. Since the sun doesn't shine 24 hours a day, you must incur the double expense of buying and installing both the solar and conventional systems. Will you save enough in utility bills to make this dual system worthwhile? At present, the answer for most homeowners is clearly in the negative.

To utilize solar heat, you must have some sort of collector, some way to store what is collected, and a way to distribute the collected and stored heat to the areas where it is needed.

The simplest and most popular collector for home solar systems is the flat plate unit. In its most basic form, the flat plate collector has three elements: (1) the transparent cover, (2) the absorber, and (3) the working fluid. The transparent cover is there to let heat from the sun radiate through to the absorber while simultaneously preventing heat loss from the absorber. The absorber can lose heat through convection (air currents carrying heat from warm to cold surfaces) and through thermal radiation (heat waves bouncing back to the sky or any cooler object). The cover also protects the absorber from precipitation and dirt, and it prevents the wind from cooling the absorber.

Glass works better than plastic as the material for the transparent cover because it stays clearer longer. For the do-it-yourselfer experimenting with a solar collector, however, transparent plastic film can serve as an inexpensive yet functional collector cover. The size of the air space between the transparent cover and the absorber is important; allow no more than about three-eighths to one-half inch since larger gaps allow too much heat loss through convection.

The absorber itself is a plate coated with a dull (matte finish) black paint or with a special black metal coating. The black color reflects less heat and aids the plate in absorbing the sun's heat. For the do-it-yourself unit, the absorber can be spray painted with a matte finish black paint. The front surface of the absorber can be flat, corrugated, or grooved, while the back surface is generally insulated to limit heat loss.

The working "fluid," usually air or water, transfers the heat from the collector to the place where it will be used or stored. It does this by picking up heat as it passes through or near the absorber and

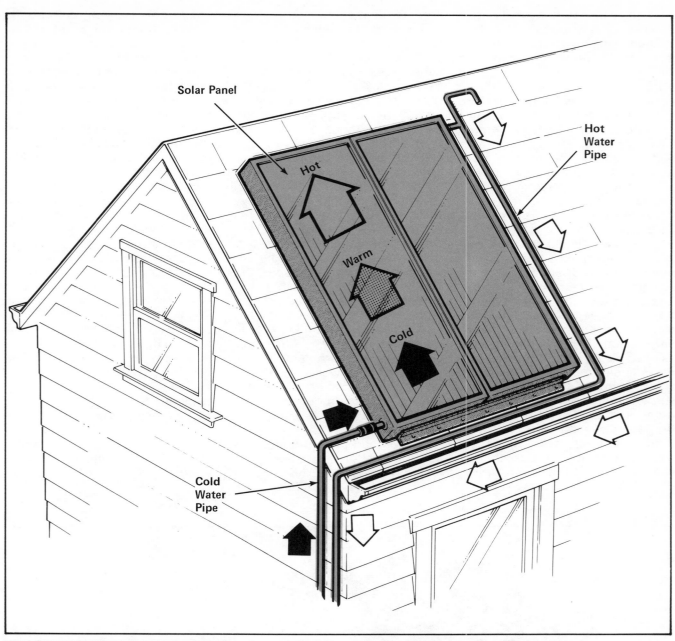

Solar Panel

Hot

Warm

Cold

Hot
Water
Pipe

Cold
Water
Pipe

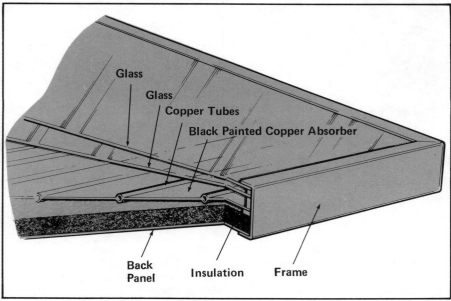

Glass

Glass

Copper Tubes

Black Painted Copper Absorber

Back
Panel

Insulation

Frame

▲
The most popular collector for home solar systems is the flat plate unit which consists of three elements: (1) the transparent cover, (2) the absorber, and (3) the working fluid. The flat plate can be incorporated easily into an existing roof, but it must be angled precisely to catch solar rays.

◄ *The absorber is coated with dull black paint or with a special black metal coating to reflect less and absorb more of the sun's heat. The front surface of the absorber can be flat, corrugated, or grooved, while the back surface is usually insulated to limit heat loss. The absorber should be quite close to the transparent cover to prevent heat loss via convection.*

then carrying the collected heat to where it is to be utilized. The medium can move naturally or be made to move with pumps or fans.

Both forms of working fluid have their disadvantages. Water must be treated with antifreeze to prevent possible freezing, while air offers a less efficient transfer of the heat to the home's hot water system and also requires an electric fan in many instances for efficient movement.

The flat plate can easily be incorporated into an existing roof, or it can be installed as a portion of a new building. Of course, the plate's tilt and orientation to the sun must be taken into consideration when determining where it is to be located. It must face the sun and be at an angle as near perpendicular to the sun's rays as possible. In the United States, the collector must generally face south, although the angle of the sun is different at different latitudes and at different times of the year. Collectors are available in fixed and adjustable versions, with some even able to track the sun automatically, and reflectors

can frequently be incorporated to increase the collector's thermal yield.

There are also collectors, called concentrating collectors, that employ curved surfaces to increase the yield of solar energy. Since concentrating collectors are much more costly than other types, however, they are not generally found in home applications.

The storage part of the system is designed to hold solar energy for use during the night or any other time when the sun's rays cannot be collected. The two main types of storage media are water and rocks. Water has the ability to hold heat well, but it suffers from two liabilities: (1) the storage tank must be protected against corrosion, and (2) large quantities of stored water present a weight problem. Rocks also hold heat well and do not lead to corrosion, but they require a larger storage space than water to retain an equivalent amount of heat.

The method of distributing the solar heat after it has been collected and stored varies accord-

ing to how the heat is to be applied. A home hot water system or pool heater has its own distribution system. In the former case, water is merely heated and then taken to a storage tank where it is available on demand. In the latter instance, the pool's pump circulates the heated water into the pool and takes water out of the pool to be circulated into the collector for heating.

Space heating distribution systems can involve duct work (for warm air) or pipes and radiators (for hot water). Some warm air systems have pumps or fans to transport heat throughout the house, while others rely on natural means such as gravity and convection to distribute the heat. Those requiring mechanical help are called active systems; those depending upon the natural distribution method are called passive.

Adding A Solar Hot Water System

THE MOST popular application—at present—of solar energy in the home involves the

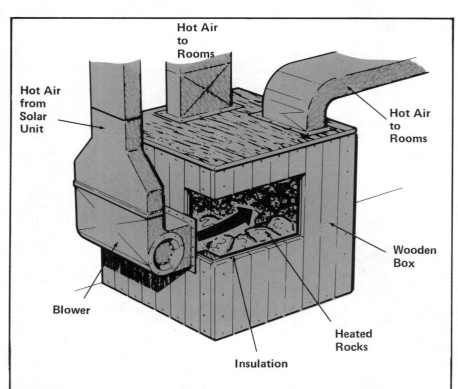

Rocks can serve as a solar system's heat storage medium. They hold heat well and do not lead to corrosion, but they require a good deal of space.

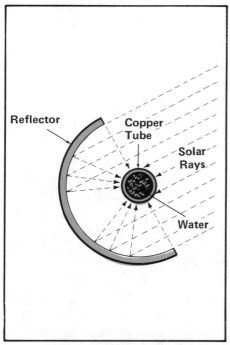

Concentrating collectors employ curved surfaces to increase the yield of solar energy. Since they are quite costly, however, concentrating collectors are seldom found in homes.

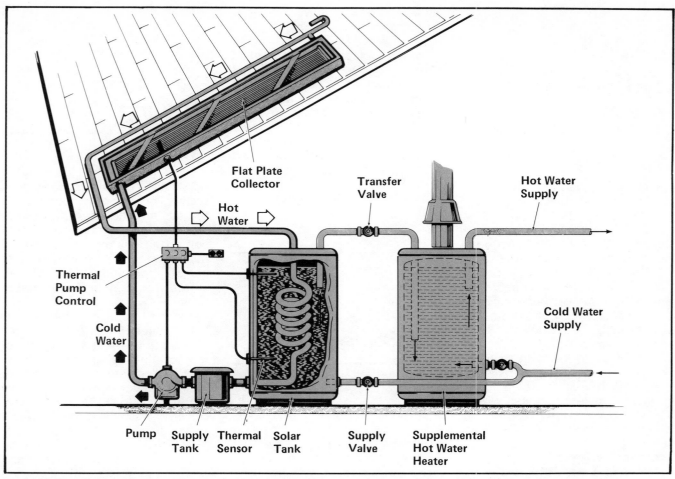

Flat Plate Collector

Transfer Valve

Hot Water Supply

Hot Water

Thermal Pump Control

Cold Water

Cold Water Supply

Pump **Supply Tank** **Thermal Sensor** **Solar Tank** **Supply Valve** **Supplemental Hot Water Heater**

In a solar hot water system, a circulator pump is generally used to carry water through the collector. The solar storage tank can be separate from the supplemental heater tank or combined with it.

hot water system. In the majority of cases, however, the solar system must be supplemented with a conventional water heater. The supplemental heater, usually electric, comes on when the sun fails to heat water to the desired temperature and during periods of peak demand.

The solar collector in a hot water system must be suitable for potable (drinking) water, and it must be constructed so as to withstand the pressure exerted by the water main. The solar storage tank can be separate from the supplemental heater tank or combined with it.

The tank(s) and the solar collector are generally connected with copper pipes, although plumbing codes in some areas permit the use of plastic pipe. A circulator pump is generally used to carry water through the collector. Each time the water goes

through the collector its temperature is increased to the maximum; thus, the desired temperature is maintained in many units by an automatic on/off switch controlling the pump. When someone turns on the hot water tap, the heated water consumed is replaced by fresh water from the supply line, which then mixes with the hot water stored in the tank and is eventually circulated through the collector.

While a solar water heater is a very simple system, it must conform strictly with the local plumbing code. The solar collector and tank should be purchased from a reputable supplier; contructing them is not a do-it-yourself project. You may well discover that the cost of these components—plus that of a supplemental standard water heater—make many solar systems uneconomical. Before you

go the solar route, therefore, add up the costs and projected energy savings to see whether such an installation would pay for itself over a reasonable length of time.

Probably the most logical do-it-yourself solar project is a swimming pool heater, although obviously the project is of interest only to pool owners. An ideal project because it does not involve potable water, the solar pool heater allows you to fabricate your own collector. In addition, the project does not involve a storage tank because the heated water is carried directly into the pool. Plumbing codes, moreover, are more lenient regarding plastic pipe for pool applications, and plastic pipe is much easier than copper pipe for the homeowner to install. Just be sure to install plastic pipe that is made for hot water use.

Hydroelectric Power

ANOTHER potential source of future energy is water power. Hoover Dam, TVA, and other enormous hydroelectric projects illustrate moving water turning a generator can create electricity. Only a small percentage of the power that water is capable of generating has yet been tapped, and, in fact, much of the potential of hydroelectric power hasn't even been measured.

In addition to the force exerted by water in rivers, streams, and waterfalls, there is a great source of water power in ocean tides. Tidal-powered hydroelectric plants are currently in existence, but only to a minor extent. Researchers are experimenting in order to find more efficient ways to utilize the constantly moving force of tidal waters.

What potential does water power offer as an alternate energy source to the typical homeowner? Not much. Even if you happen to live on a river or stream, water-rights laws or environmental safeguards may well preclude your being able to take advantage of the water as a power source. You would also have to overcome the problem of inconsistency in using water as a water source. The force of the water may vary from season to season, with some streams completely drying up at certain times of the year. In most cases, therefore, water power cannot be depended upon for a total year-round source of electrical supply.

If you need power in a remote area where utility lines are not available, however, a hydroelectric system would be worth considering as an alternative to a gasoline-powered generator. The initial cost would be greater, but over a period of years the hydroelectric system would provide a good return on your investment.

As energy demands rise and as new hydroelectric equipment is developed, the time may well come when water power will become a plausible alternate source of energy for many American homeowners. Until then, hydroelectric power will remain an interesting subject to know about and perhaps pursue as a supplemental power source where circumstances permit.

Methane Gas

EVERYONE knows that it takes energy to dispose of garbage, but not everyone is aware that garbage itself may be the source of a great deal of critically needed power. We are a nation that creates tons of garbage every minute, and much of this garbage can be transformed into methane gas. Organic material that decomposes rapidly—e.g., food scraps; animal and human waste; plant clippings, weeds, and leaves—can be used to produce methane gas.

In the process of making the gas, moreover, the basic organic material is composted into excellent fertilizer. Compost can replace some of the chemical fertilizers made from petroleum byproducts, making the waste material even more valuable.

The units designed to extract gas from waste material are called digesters, and the biggest problem at present is that the digesters currently available require huge amounts of waste to create a usable amount of methane gas. As experimenters continue to find new methods of garbage digestion, however, common household waste may well become an important—albeit partial—solution to the current energy crisis.

Sky Therm Cooling

A TOTALLY natural process that collects heat from the sun during the day and holds it until night, the sky therm cooling system cools by radiating the collected heat into the night sky. In its most basic application, the system stores the heat in an insulated pool of water, preventing the transfer of heat into the interior living space. Since the water (or other liquid) must be on top of the building being cooled, however, only existing buildings with flat roofs can be economically equipped with sky therm cooling. Nevertheless, the system clearly works in principle, and appropriate technology may soon be developed to permit widespread consumer applications.

Wind Power

AT ONE TIME, windmills were commonly found on farms providing the energy needed for pumping water, grinding grain, and other mechanical chores. But rural electrification did away with the need for wind power until the present energy shortage. Now, though, many people are talking about moving back to wind power to drive electric generators.

While the wind power itself is free, converting a puff to a watt can be quite expensive. More than just a propeller spinning in

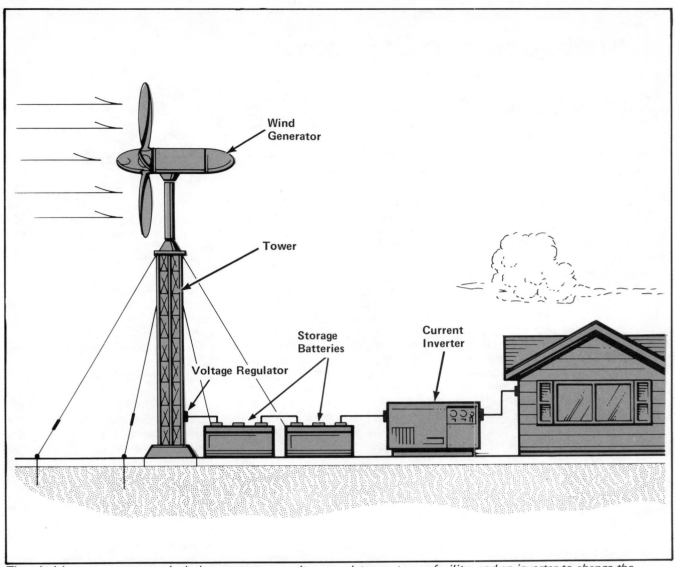

Wind
Generator

Tower

Storage
Batteries

Current
Inverter

Voltage Regulator

The windcharger system must include a generator, a voltage regulator, a storage facility, and an inverter to change the current from direct (DC) to alternating (AC). Before you invest in this large and costly equipment, you must determine whether your site offers enough wind to make the system function. Most wind power systems reach their rated output with winds of about 20 mph.

the breeze, the windcharger system must include a generator, a voltage regulator, a storage facility, and—if the electricity is consumed in a typical home—an inverter to change the current from direct (DC) to alternating (AC). Consequently, to produce enough power for the average home, a good deal of large and costly equipment is required.

Remember, too, that even with an expensive windcharger system no wind means no power. Storage batteries can tide you over a temporary lull, but to make a wind power system economically feasible, you must live in an area that provides a fairly con-

stant minimum wind velocity of ten to twenty miles per hour. Most systems reach their rated output with winds around twenty miles per hour, although there are a few that will start generating power when driven by as little as seven-mph winds.

Before you even consider looking into a windcharger, therefore, you must determine whether your site offers enough wind to make the system function. You also must make certain that the area around where you plan to put up your windcharger will not be surrounded in the future by tall buildings or anything else that would block the wind.

Thus, as a means of power for most homes, wind systems must be entered on the list of impractical alternate energy sources. There are areas, however, where despite its high cost and obvious disadvantages wind power is used because it is the best available source of power. If you are one of those people who needs a power source where conventional power isn't available, or if you want to supplement your conventional electrical supply, or if you want to put up a windmill to pump water, you'll be happy to know that all the components needed are available to anyone who wishes to buy them.

Alternate Sources Of Energy

Materials and Supplies

Lennox Industries Inc. now markets a complete solar domestic hot water system that can supply between 50 to 100 percent of the hot water used by a home or small business. The system, known as the LSHW1 series, consists of high-efficiency Lennox flat-plate solar collectors and a Lennox solar hot water storage module available in 66-, 82- and 120-gallon capacities. The system can be installed in new construction or put in as a replacement hot water system.(1)

Sun-powered water heaters for homes and apartments are being produced by State Industries. The Solarcraft system, designed to provide from 55 to 85 percent of a family's hot water needs, includes an 82- or 120-gallon hot water storage tank, roof-mounted solar panels, controls and a fluid handling system. A mid-tank auxilliary heating element assures hot water on cloudy days.

Among those companies marketing residential domestic hot water systems is American Heliothermal Corporation. The AHC hot water system is of the closed loop variety, which means that after the working fluid is heated, it goes to a storage tank where its heat is transferred to another tank containing the actual water to be consumed. The advantage to this type of system is that antifreeze can be added to the working fluid to prevent freezing. AHC offers a supplementary system in the form of an electric heating element or a separate heater.

Mueller Climatrol Corporation has developed a solar-assisted heat pump system that combines solar hydronic heating with an energy-efficient heat pump. The system consists of a Climatrol Climator II split-system heat pump, a bank of solar panels, a thermal storage tank, a duct-mounted solar hydronic heating coil and a control panel. A black

absorber plate located in each solar panel collects energy from the sun and uses it to heat water in the storage tank. On moderately cool days, the system actually functions as a conventional fan-coil hot water heating system. When the temperature drops, heating operation is automatically switched to the heat pump. Additional electric heat can be activated by an outdoor thermostat as required.(2)

Solar Works Inc. has developed the Solar Wunder portable solar water heater. Perhaps the most innovative idea yet for the practical use of solar energy, it can heat water, coffee, soup, hot chocolate, or other hot beverages in just one to two hours; it can also boil water if left in the sunlight for a longer period of time. The unit is made of lightweight durable plastics, is fully thermal insulated, measures 25x10x9 inches and, when filled to its two-gallon capacity, weighs

just 25 pounds. It requires no fuel, no batteries and has no moving parts of any kind. The Solar Wunder stores hot water or beverages all day or overnight, retaining all but a few degrees of its stored heat. (4)

The Solaire 36 is a three-ton residential sun-powered air conditioning system from Arkla Industries, Inc. More compact in size than previous models, the Solaire 36 also features more flexibility in both installation and operation.

If you have your own swimming pool, you may want to invest in a solar pool heater. Depending upon location, these heaters can be totally effective, requiring no auxiliary heat sources. Fafco pool heaters utilize lightweight plastic modular units that can be formed to create any size collector panel needed. An automatic control turns the system on when the sun is hot enough to heat the pool water and off at other times.

The Solar Furnace made by the Solar Division of Champion Home Builders Company can be connected to the duct work of an existing forced air system. The conventional unit becomes the back-up system for the Solar Furnace. Since the collectors are ground type, no structural changes are required in adding the Solar Furnace to a forced air system. Storage is accomplished by using screened and washed gravel. The Solar Furnace comes in three sizes.

Solar Shelter's forced air solar furnace can be added to most existing houses. In addition to ground level collectors, aluminum reflector plates and a brick pile storage area, the unit employs two blowers. One circulates the heated air from the collectors, while another carries heated air into the house on demand from the thermostat. When heat in the brick pile drops below a certain level, the auxiliary heating system kicks in. Once installed, the solar furnace is practically maintenance-free; only the fan belts would require possible replacement. (5)

The Sunmat from Calmac Manufacturing Corporation is just one of the many separate solar collectors presently available. They come in a multitude of styles and sizes, and with a wide variety of prices too. Other makers of flat plate collectors include Sunearth Inc., Energy Systems Inc., Hitachi, Columbia, American Heliothermal Corp., Sun Sponge, TechniTrek Inc., Libbey-Owens-Ford Co., Olin Brass, General Electric, New Jersey Aluminum, Reynolds Aluminum, Solar Components Division of Kalwall, American Solar King Corp., Solar Development Inc., and Revere. Although many share the same basic principles, several unique features distinguish a few of these brands. (3)

The PPG domestic hot water system uses tap water as the medium for solar heating. Storage tanks come in several sizes, and the system uses flat plate collectors—the number of which depends on the area in which the unit will operate. Installation normally takes about a day and a half and can be done by a plumber or competent homeowner.

The ability of most solar devices to perform at their best is usually dependent on a special control that compares the temperature of the medium in the solar collector with that in the storage tank. The control activates the circulator that moves the solar-heated water down to the storage tank and moves water from the bottom of the tank back through the collector. The control shuts off the circulator when no solar heating can occur because the circulation of the water from the tank could then result in cooling. These control devices also limit temperatures from going too high, and they protect the heating system from freezing. Rho Sigma Inc. makes a number of control units for solar applications, including one for use with water heating systems.

(1) Lennox LSHW1 series hot water system. (2) Mueller Climatrol's heat pump system includes banks of solar panels. (3) Sunmat solar collector. (4) Solar Wunder portable water heater. (5) Solar Shelter forced air furnace includes ground level solar collectors.

145

Lighting

ALTHOUGH MANY electrical appliances consume far more electricity than does a single light bulb, lighting still accounts for approximately 22 percent of the total electricity a typical family of four uses. In most cases, a family can cut back on this electrical consumption without sacrificing the desired level of lighting and without incurring any discomfort or danger.

Bulbs

THE ONE figure you examine when installing a new bulb is its wattage. Wattage is a measure of the energy the bulb uses, and most people operate under the theory—which is valid in most cases—that when more light is needed, a bulb with a higher wattage should be installed.

The true measure of the amount of light a bulb produces, however, is its lumens. A bulb that provides greater lumens while using fewer watts is, therefore, an energy saver. Check to see how many lumens a bulb puts out. The lumens measurement is required by law for general service incandescents and is usually indicated for other types of bulbs as well. You may find the figures on the bulb itself, but usually the lumens can be found on the bulb's container.

So-called long-life bulbs put out less light (smaller lumens measurement) than a standard bulb of the same wattage. The standard bulb may last a shorter time, but it produces more brightness while in use. Thus, a standard bulb and a long-life bulb of the same wattage consume the same amount of energy over their life times. Where long-life bulbs are valuable is not where energy saving is a consideration but rather where frequent bulb changing is a nuisance.

Incandescent bulbs of smaller wattages are generally less efficient than are similar bulbs of larger wattages. For example, a standard 100-watt bulb provides more light while consuming less energy than do two standard 60-watt bulbs.

Fluorescent tubes develop light much more efficiently than do incandescent bulbs (the ones with filament wires that light up when current passes through them). A 40-watt fluorescent, for example, gives more light while using less than half the current of a 100-watt incandescent. In addition, fluorescent tubes emit much less heat, an important advantage during the summer. Finally, fluorescent tubes outlast incandescent bulbs, usually by about ten times and sometimes up to 20 times.

Why not convert all incandescent bulbs to fluorescents? The problem is that fluorescents require special fixtures, and it is quite costly to make a complete changeover from one system to the other. You certainly should consider switching to fluorescents, however, every time you add new lighting or replace an existing fixture.

Another type of high-efficiency bulb is the HID (high intensity discharge). While HID fixtures and bulbs are not applicable to all lighting situations, they can be used in small areas requiring highly intensified light beams. The HID bulb provides two to five times the light of an incandescent bulb of the same wattage and can last from ten to 25 times longer.

Lighting Tips

HERE ARE some ways to save energy when it comes to artificial lighting in your home.

1. Turn off all unneeded lights. Turn off incandescents every time you leave the room—if only for a few minutes. Fluorescents, on the other hand, should be left on if they will only be off for 15 minutes or less. Fluorescents need more electrical energy to get started, and frequent starts reduce tube life.

2. Convert incandescent to fluorescent fixtures where practical.

3. Install dimmer switches. A dimmer (interchangeable with a regular wall switch and wired in the same way) lets you turn down the wattage when less light is required. If the switch also controls convenience outlets where an appliance can be plugged in, however, don't install a dimmer.

4. Use three-way bulbs in lamps when possible. A three-way bulb allows you to adjust the lighting intensity to the lighting needs at any given time.

5. Watch for darkened bulbs. They give off less light.

6. Keep all bulbs and tubes clean. A dusty bulb sheds less light.

7. Reduce the size (wattage) of bulbs where only fill-in or general light is needed.

8. If you install bright security lights, connect a photoelectric cell or timer to turn the lights on at dusk and off at dawn.

9. Use windows and other sources of natural light instead of artificial lighting as frequently as possible. Keep the windows clean, and keep drapes and blinds open unless heat absorption is a problem in summer.

10. Check to make sure that all the bulbs in remote places—e.g., the attic, basement, garage, closets, and tool shed—haven't been left burning. Automatic switches in closets that shut off the light when the closet door is closed can be a good investment.

11. Use spotlight bulbs over areas where you need good light for a specific task. Reflective bulbs and fixtures intensify the light where needed.

12. Use colors to help reflect the available light.

Fluorescent Fixtures

THERE ARE three basic types of fluorescents: the pre-heat, the rapid start, and the instant start. Pre-heat units use a starter mechanism, while the other two types do not.

Every fluorescent unit, though, contains a component called a ballast, a sort of transformer that converts house current into a current of its own that produces light. The type of ballast in a fluorescent fixture is also your guide to purchasing all replacement components—tubes, starters, or the ballast itself. You must install the exact replacement parts or you will never enjoy properly working fluorescent light fixtures.

If you get no light at all from your fluorescent fixture, follow these basic steps.

Step 1. Make sure that the house current is on and that the circuit is live by checking the fuse or circuit breaker box.

Step 2. Next, make sure that the fluorescent tube is inserted properly in the lamp holder.

Step 3. If you have another fluorescent fixture that is working, try the nonfunctioning tube in the good fixture to see whether the tube is burned out.

Step 4. If the tube is still good, replace the starter—provided you have a pre-heat fluorescent fixture. The starter is an inexpensive and easy component

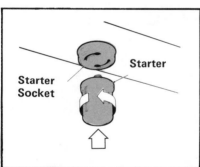 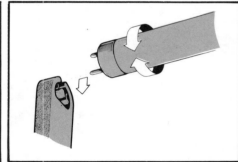

If a fluorescent fixture fails to function and you have reason to believe that the tube is still good, check to make sure that the tube is inserted properly (right), and then replace the starter mechanism (left).

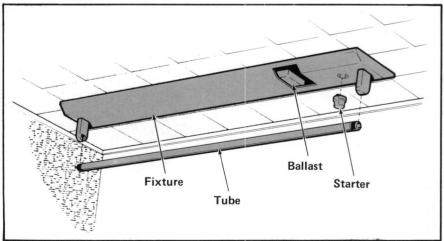

A pre-heat fluorescent fixture consists of three primary components, the tube, the starter, and the ballast. The type of ballast in the fixture is a guide to obtaining replacement parts for the other components.

to replace. Just be sure to buy the right replacement for your fixture.

Step 5. If a new starter fails to make the light work, replace the tube. Again, be sure to get the proper size.

If the light blinks off and on, here is the procedure to follow:

Step 1. Make sure that the temperature in the room is more than 65 degrees Fahrenheit.

Step 2. Check to be sure that the tube is seated firmly in the lamp holder.

Step 3. Inspect the pins on the ends of the tube to see that they are free of dirt or corrosion. If they are not, sand them lightly. If the pins are bent, use a pair of pliers to straighten them gently.

Step 4. Cut the current to the fixture at the fuse or circuit breaker box, and inspect the wiring for loose connections.

Step 5. If none of the above steps stop the light from blinking off and on, replace the starter and then the tube.

If the tube is dark in the middle but the ends light up, or if the ends are discolored to more than just a brownish tinge, these steps should restore the lamp to proper working order.

Step 1. Be sure that the room temperature is high enough. It should be 65 degrees or higher.

Step 2. Replace the starter.

Step 3. Remove the tube and reverse it—end for end—in the holder.

Step 4. Cut off the circuit that supplies current to the fixture, and check the wiring inside the unit.

If there is a pronounced flicker in the light, try these basic steps to get it working right.

Step 1. If it is a new tube, the flicker will go away soon. New tubes often flicker at first.

Step 2. Shut the fixture off and then turn it back on.

Step 3. Replace the starter.

Step 4. Cut off the current and check the wiring inside the fixture.

If you experience any problems right after you replace parts, check the legend on the ballast to verify that the parts are the right ones for the fixture.

Natural Lighting

ONE WAY to cut lighting costs is to utilize free lighting from the sun whenever possible. To do this, of course, means creating openings through which sunlight can enter. Unfortunately, it is not often possible to control the direction of this natural light, and if you want to make good use of it, you may have to move your furniture around. Reading, writing, and sewing require a good deal of lighting, and you should arrange your furniture so that the sewing machine and/or desk receives the natural sunlight needed to work comfortably.

If you're thinking about adding windows or building an addition, first consider the heat transfer involved. Make sure that you understand all the ways to cut down on heat transfer through glass; otherwise, you could find that your new windows create a major energy loss. Carefully planned windows, however, will be an energy asset.

Since the idea of adding windows is to take advantage of the natural light year-round while minimizing heat gain during the summer, avoid large openings on the west and north. Instead, put windows on the eastern and southern exposures. That way, you'll pick up some heat from the low winter sun while avoiding the worst of the high summer sun.

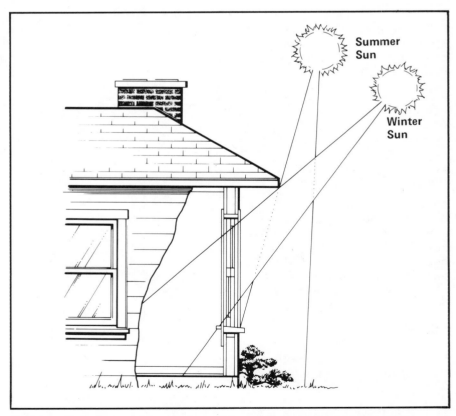

Position windows to pick up the low winter sun but not the high summer sun.

Skylights

SKYLIGHTS represent a special way to admit natural light into your home. For some rooms that are in the center of the house, moreover, skylights may be the only way to admit any natural light.

An interior bathroom or other inside room can often benefit from a skylight in terms of both light and ventilation. All interior rooms should have some means of ventilation, and a proper skylight may be the best way to achieve such air circulation. In addition, a skylight can also be important in moisture removal from a bathroom.

Skylights are available in pre-assembled units, or they can be custom-made to suit your needs. The factory-made units come in a variety of sizes and styles, including the bubble type which consists of a formed plastic bubble that is either clear, tinted, or frosted. Other factory-made skylights have flat panes of either plastic or glass. The better flat types are multi-glazed units.

Skylights come in fixed versions and in operable units that can be opened and closed. Where ventilation and moisture removal are important, opt for an operable skylight.

Many skylights, of both clear and tinted varieties, employ a frosted diffuser to spread the light over a wider area. Either the prime glazing material is frosted or else a second pane of frosted material is added below the prime one. Such diffusion is not necessary, of course, if the skylight is installed only to increase the light in a small inside bathroom, closet, or other specific spot of limited dimensions.

Naturally, there are drawbacks to skylights. Since a skylight is located at the top of the room, it may lead to more heat being lost

The bubble type of skylight is available in pre-assembled units, consisting of a formed plastic bubble that is either clear, tinted, or frosted.

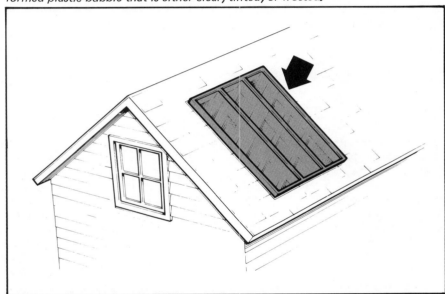

Other factory-made skylights feature flat panes of either plastic or glass. The better versions are multi-glazed and include a frosted defuser to spread the light over a wide area of the room below.

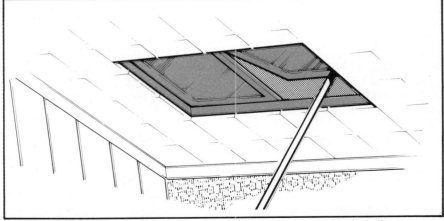

Where ventilation and moisture removal are important concerns, install an operable skylight that can be opened and closed.

149

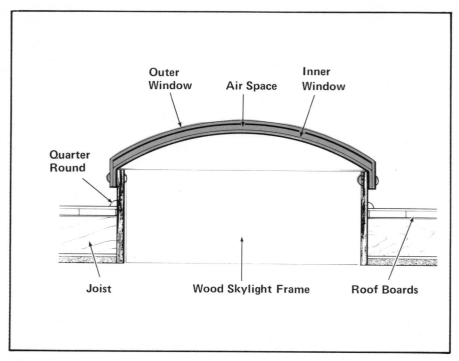

Multi-glazing, a standard feature of the better factory-made skylights, can reduce heat loss substantially. If heat loss is a problem during long winter nights, insulate the skylight at the bottom of the light shaft.

during long winter nights than is gained during short winter days. Conversely, solar heat gain—an asset during the winter—can be a disadvantage during summer.

Heat loss can be reduced through multi-glazing, a standard feature of the better factory-made units. In addition, a homeowner can slip an insulator panel—a scrap piece of suspended flexible ceiling material will do the job and look fine —into the opening at the bottom of the light shaft at night and remove it during the day. Plastic diffuser panels also help along these lines.

The easiest factory-made skylight for a homeowner to install is the type that fits between two adjoining rafters. Although this type still requires that you cut through the roof, it saves you from having to cut through the rafters as well and then compensating for the structural strength lost. Larger factory-made units require removal of part of the rafter and putting a header in to replace the strength lost. Although this carpentry chore is not difficult, it must be done according to the

provisions of local building codes. Factory-made units are usually made of materials that are in conformity with code requirements, and they are usually quite easy to seal against leaks.

If you wish to fashion your own skylight, be sure that your design and materials are code acceptable. Many building codes require that such skylights possess wire-reinforced glass.

Paint And Color Reflective Factors

SINCE DARK colors absorb light while light colors reflect it, the proper use of reflectance can greatly increase the efficiency of your lighting. If you know the reflectance values of different col-

ors, you'll be able to take better advantage of reflected light the next time you paint.

Keep in mind that ceilings are the most important surfaces for maximizing lighting efficiency through reflectance. Even if you do not care for a light color on walls, a light color on the ceiling will not mar the room's aesthetics while assisting significantly in its lighting properties.

In areas of your home where visual demands (e.g., reading, sewing, etc.) are minimal, you can use any colors that you wish. A hallway, for example, needs no reflective light; a study, on the other hand, should be painted in a color possessing a high reflectance percentage.

The percentages in the following table represent the reflectance values of various paint colors. Since these values can vary according to the paint manufacturer, however, they must be regarded as approximate figures. The percentages are for nonglossy wall and ceiling paints which diffuse light better than do glossy versions of the same colors; the glossy finishes, on the other hand, possess slightly higher reflection percentages.

Paint Color	Reflectance Percentage
White	82%
Ivory	78%
Yellow	75%
Flesh	71%
Light Pink	70%
Beige	69%
Light Gray	65%
Lemon Yellow	64%
Powder Blue	63%
Aqua	61%
Apricot	59%
Gold	54%
Rose	46%
Medium Gray	45%
Orange	35%
Coffee	28%
Kelly Green	25%
Red	21%
Chalk Board Green	20%
Brown	15%
Walnut Paneling	10%
Dark Royal BLue	08%

Lighting

Materials and Supplies

Fluorescent Fixtures

You can easily convert a regular ceiling fixture from incandescent to fluorescent with the Conserve-A-Lite from the American Fluorescent Corporation. The unit consists of a round 22-watt tube connected to either a chrome or white base. Although not an aesthetically appealing lighting fixture, the Conserve-A-Lite is well suited to installation in a workshop, basement, garage or utility room. Best of all, no wiring is required; the Conserve-A-Lite base merely screws into a standard incandescent socket.

The Killer Watt from Johnson Industries is a screw-in adapter that goes into any ordinary incandescent light socket and converts the fixture to fluorescent. Available in either single or double tube units, the Killer Watt functions equally well in lamps and ceiling fixtures. (1)

General Electric's Bright Stik makes conversion to fluorescent lighting easier than ever. The unit is its own fixture and therefore can be installed anywhere its electrical cord can be plugged in. On the back of the Bright Stik are self-adhesive pads for positioning the fixture, screws are then driven for a quick permanent installation. The 33-watt unit measures only 25 inches long. (2)

Solite makes two fluorescent devices designed to reduce home lighting costs. Tight Watt is an energy-saving fluorescent fixture that screws into a light bulb socket, table lamp, or into a junction box for a direct ceiling mount. The Lite Box, a ceiling

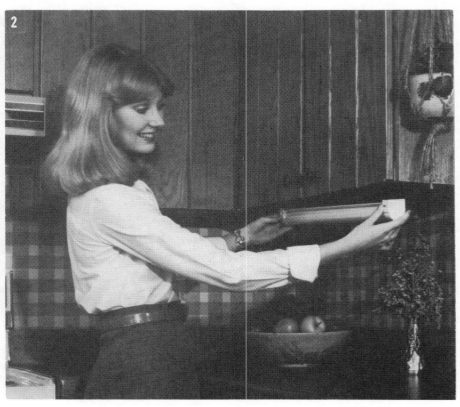

(1) Killer Watt screw-in adapter. (2) Bright Stik fluorescent fixture.

type fluorescent lighting fixture, must be wired into place. Available with two or four lamps, the Lite Box offers substantial energy savings while providing more light than comparable incandescent fixtures.

The most critical component of a fluorescent fixture is a black box called the ballast. When the ballast goes on the blink, fluorescent fixtures fail to function. Finding the right replacement ballast used to be a problem, but the Handyman Electric Company has developed a line of ballast units that will replace all of the standard ballasts. All you need to know is the manufacturer's name and model number of your defective ballast to find a Handyman replacement. Instructions for installing the new ballast come on the back of the package; the homeowner can certainly handle the job.

Dimmer Switches

Dimmer switches come in two versions. One has two settings, bright and dim, while the other has a continuous dial that lets you adjust the light intensity all the way from bright to off. Hemco makes moderately priced dimmers of both types; either type will fit in the space now occupied by a conventional wall switch, and installation is quite simple. (3)

The Premier, GE's top-of-the-line dimmer switch, has a lighted dial that can be adjusted to any degree of brightness. Features include a push switch, advanced high attenuation filtering of radio interference, 600-watt capacity, and enhanced surge resistance for cooler operation. The Premier can be wired to replace either a standard switch or a three-way switch.(4)

You can convert an ordinary table lamp into a lamp with a full-range dimmer by merely plugging the lamp into the Leviton Table-top Dimmer. The Leviton unit requires no electrical wiring whatsoever. You merely plug the lamp into the dimmer; the dimmer into a wall outlet. (5)

(3) Hemco light dimmer switches. (4) Premier adjustable dimmer switch. (5) Leviton Table-Top plug-in dimmer unit.

Energy Emergencies

EVEN WHEN energy itself is not in short supply, emergencies can occur that knock out the service from utility companies. And the prospect of genuine and prolonged fuel shortages appears more threatening every day. While everyone hopes never to face such problems, prudent homeowners are learning certain procedures designed to make life more bearable and safer during such stoppages.

Electrical Blackouts

IF YOU'RE watching TV and suddenly the set goes dark at the same time the lights go out, you know that there has been an electrical power failure. The first thing to do is look out the window to see if lights are out all over the neighborhood. If you see lights nearby, you know that the trouble is probably at your fuse box or circuit breaker box, and you should take steps to find and correct the problem. If there are no lights visible, however, there is little you can do to solve the problem. But there are steps you can take to cope with the situation.

It's a good idea to have several flashlights in good working order around your home in case of a power failure. Put them where they are easy to find. You should have one at the fuse box, one at the head and foot of all stairways, and others where you think they would be of the most help. Candles can also come in handy, and camp lanterns are great.

If you live in an area that suffers blackouts frequently, you should invest in one or more of the emergency lights that come on automatically whenever the current goes off. You leave these units plugged into wall outlets, and when they no longer receive electricity, they light up.

As soon as you have emergency lighting, you should go around and turn off all appliances and other electrically powered devices. You don't want to jolt your own electrical system when the power is restored. Once power is restored, you can go around and turn all these devices back on one by one.

Since the cold air stored in your refrigerator and freezer will keep foods from spoiling for several hours, you should open these appliances only when you absolutely must. A freezer that is kept closed can prevent spoilage for 36 hours or more.

During the winter, an electrical blackout can cause heating problems even in homes that don't depend on electricity for heating. Most gas and oil furnaces have controls that operate on electricity, and when the electricity is cut off, the furnace will not function.

If you live in an area of frequent blackouts, ask the local power company for information regarding how you can safely activate your particular type of furnace manually and what precautions you must take. Never attempt to bypass the automatic controls without knowing what you're doing. The same controls that turn the furnace on also act as safety devices, shutting the unit off before the heat reaches a dangerous level; they also control the fan that carries the heat away.

A fireplace can be a good source of emergency heat, although its effect is limited to one room. A gas stove will warm the kitchen, and if you keep a pan of water boiling, the steam will carry warmth beyond the kitchen area.

Along similar lines, if you have a gas water heater you can fill bathtubs and sinks with hot water. Much of the heat will reach other rooms before the water cools.

If you face the prospect of frequent power blackouts, however, you might wish to consider having your own electrical plant—i.e., a gasoline generator. Although an impractical alternative source of everyday energy due to the oil situation, the gasoline generator can nevertheless

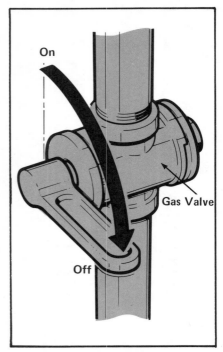

When a natural gas stoppage strikes, shut off all gas supply lines to the furnace, space heater, water heater, and stove. Although most of these units have built-in safety devices to prevent problems when the gas supply is restored, you should turn the gas back on only after you relight the pilot lights.

come in very handy as a standby system to power your home during electrical failures. Install a gasoline generator and you need not worry about food spoiling from lack of refrigeration, frozen pipes bursting and damaging your property, and other assorted ills that can strike when the power is out.

Gasoline generators made for consumers are compact and require no new wiring within your home. Since they are generally connected to your electrical system at the meter, however, you must consult your local power company and comply with all the regulations in your local building code—which probably means having a licensed electrician make the hookup.

Some generators automatically switch on whenever there is a stoppage of power from the utility company, while other units are portable and need only be plugged into specially installed connector switches when standby power is needed.

Natural Gas Stoppages

NO GAS usually means no heat and no hot water, and it can also mean no cooking. Alternate sources of room heat include the fireplace and electrical space heaters, while cooking can be done with electric hot plates, fry pans, and other small electrical cooking appliances.

When a natural gas stoppage strikes, you must take certain emergency actions. You must shut off all gas supply lines to the furnace, space heater, water heater, and stove. While most of the pilots on these units have built-in safety devices that prevent problems when the gas supply is restored, you should turn the gas cutoffs back on one by one as you relight the pilots. Make certain that any pilotless space heaters are turned to their "off" position before service is restored.

Water Supply Emergencies

MOST OF THE time, you know in advance when your water will be cut off for any extended period. Service personnel making plumbing repairs or repairs to other utilities that necessitate cutting the water supply will inform you of the situation and give you time to store water for drinking and cooking.

If you face an unexpected emergency, though, you should

know a few steps you can take to get through the shortage. Insofar as drinking water is concerned, look to the ice in your freezer as your first source. In addition, you can drink the liquid in which many fruits and vegetables are packed when canned. During the winter, of course, you can melt any snow that's around.

Another source that you might not think of is the water in the toilet tank. It comes from the same supply lines as your drinking water, and it is still potable as long as it's in the tank. On the other hand, never drink any water without first knowing that it is all right following natural disasters that disrupt normal water supply lines.

If the water is off long enough that your water heater becomes emptied, you must turn off the heat source—be it gas or electric—until new water enters the tank.

Plumbing Emergencies

IF A PIPE should burst or a joint come loose in your plumbing system, thousands of gallons of water could escape, damaging your home and furnishings. You cannot wait for someone to come out and repair the pipe or joint. Instead, you must shut off the main water supply immediately.

Since everyone should know how to shut off the flow of water, take the entire family on a tour of your home to see where the cutoffs are located. Each sink, basin, tub, and appliance should have its own cutoff, but you will find that many do not. Everyone must, therefore, know the location of the main water cutoff.

Usually located very close to where the water supply enters the house, the main cutoff may be a gate valve or it may be an

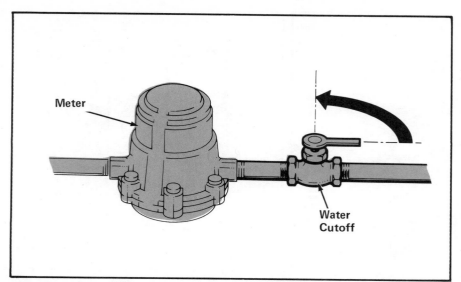

The main water cutoff is usually located very close to where the water supply enters the house. Find out exactly how the cutoff works and check its operation. Valves can corrode and become impossible to close.

"L"-shaped rod. If the meter is in the basement, you could have a meter valve that requires a wrench to turn. An underground meter would have the same type of valve, but a buried pipe means that you must dig to find the cutoff.

Once you locate the main cutoff valve, find out exactly how it works and, most importantly, if it works. Little-used valves have a way of corroding, and it is better to find out now that the valve

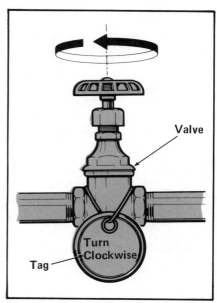

Know which way to turn the valve to shut off the water. A tag on the valve can be helpful in an emergency.

needs replacement than when you are struggling in a panic situation. In the case of the "L"-shaped rod, the actual valve is usually underground. If the valve malfunctions, you must dig down and treat the valve with penetrating oil to free it.

Make sure that you know which way to turn the valve to shut off the water. It might even be a good idea to tag the cutoff with a drawing that indicates what the valve is and which way to turn it. You can use the occasion to acquaint the family with gas and electricity cutoffs as well. As for preventive maintenance, it is a good idea to turn the water cutoff valve off and then on again once every six months or so to keep it in working order.

One word of caution: Remember that the flooding waters of a plumbing emergency can come in contact with electrical components. Make sure that you do not complete the fatal triangle. If there is any chance of electrical contact, always throw the main switch to your household current before you go wading around in the water.

If the flood emanates from a leaky tank, try one of the self-tapping plugs that are designed to go into the hole. When tightened down, the screw applies enough pressure to stop the

leak. If the tank rotted through in one place, however, it may well do so in another. On the other hand, the plugs may buy you enough time to shop around for the best replacement tank while preventing damage to your home.

The most important thing to remember during a plumbing emergency is to keep a cool head. Stop the flow of rising water to minimize damage. Once you get the water shut off, there is no reason to panic; you even have time to figure out your next move. You may decide that the time has come to call in professional help, but you may conclude that you can correct the problem yourself.

When a pipe bursts, for example, it can do tremendous damage to your home. But if you act quickly and follow these steps, you can turn a potential disaster into an easy do-it-yourself repair job.

Step 1. Cut off the water. If there is a cutoff for just that section of pipe, shut it off. If not, go to the main water cutoff for your home and turn it off.

Step 2. Locate and examine the leaky section.

Step 3. If the leak is in the middle of a length of galvanized pipe, there are several ways to clamp a patch over the hole. You can buy a kit at the hardware store that includes a pair of clamps, a rubber pad, and bolts. Or you can apply the same sort of patch with a scrap of inner tube and a worm gear hose clamp.

Step 4. For a temporary repair of a small leak, you can wrap waterproof tape around any type of pipe. The tape should cover several inches to either side of the hole.

Step 5. You can apply epoxy metal to any metal pipe. Be sure to allow the pipe to dry completely, and then follow the directions for curing time.

Step 6. Leaks around pipe joints are more common than pipes bursting. Galvanized pipe joints are threaded, and sometimes

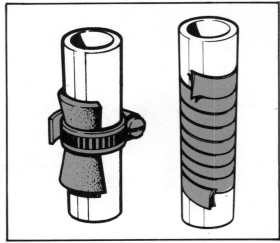

You can patch small leaks in galvanized pipe by attaching a scrap of rubber and a clamp (left) or by wrapping the pipe with waterproof tape (right).

◄You can buy a pipe patch kit at the hardware store that consists of a pair of clamps, a rubber pad, and several bolts. Just put the pad over the leak and then tighten the clamps to hold the pad securely over the faulty section.

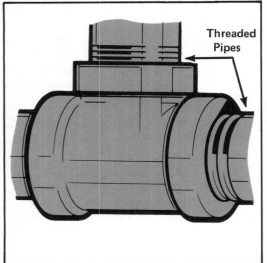

Threaded Pipes

▲ Leaks around threaded pipe joints are far more common than a pipe bursting.

Threaded galvanized pipe joints that leak can ▶ often be repaired merely by tightening. But if that doesn't work, loosen the fitting, apply pipe joint compound, and retighten.

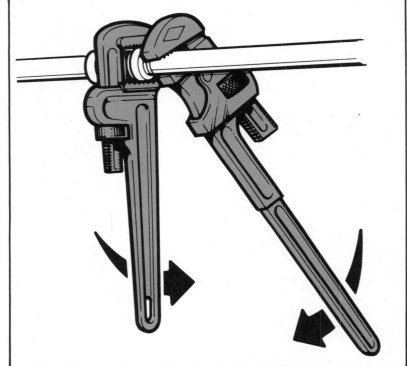

they just need to be tightened. If tightening fails to stop the leak, loosen the fitting, apply pipe joint compound, and retighten. A compression fitting that leaks probably just needs a slight tightening; but joints in copper tubing that leak must be removed, all the old solder cleaned away, the tubing dried completely, and then the joint resoldered.

Step 7. A bad section of pipe should be replaced rather than patched. Cut the bad part out, and replace it with two shorter pipe sections joined by a fitting called a union.

Step 8. If the leak is in plastic pipe, try to use solvent weld compound to seal the leak. If the compound does not do the job, replace the bad section of pipe.

Directory of Manufacturers

A

Accel Eliminator Ignition
Box 142
Bronford, CT 06405

AC Spark Plug Div.
General Motors Corp.
1300 N. Dort Hwy.
Flint, MI 48556

Admiral Corp.
Rockwell International
1701 E. Woodfield Rd.
Schaumburg, IL 60172

AGA Corp.
550 County Ave.
Secaucus, NJ 07094

Albion Engineering Co.
1572 Adams Ave.
Philadelphia, PA 19124

Allied Plastics, Inc.
1663 Hargrove Ave.
Gastonia, NC 28052

Alsco Anaconda
1 Cascade Plaza
Akron, OH 44308

Amana Refrigeration, Inc.
Amana, IA 52203

American Air Filter Co., Inc.
P.O. Box 21127, Standiford Station
Louisville, KY 40221

American Fluorescent Corp.
238 N. Oakley Blvd.
Chicago, IL 60612

American Grease Stick Co.
2651 Hoyt
Muskegon, MI 49443

American Heliothermal Corp.
2625 S. Santa Fe Dr.
Denver, CO 80223

American Standard Mfg. Co.
3801 S. Ashland Ave.
Chicago, IL 60609

AMF/Paragon Electric Co.
606 Parkway Blvd.
Two Rivers, WI 54241

Ammark Corp.
12-22 River Rd.
Fair Lawn, NJ 07410

Amspec, Inc.
1880 Mackenzie Dr.
Columbus, OH 43220

Aquappliances, Inc.
135 Sunshine Ln.
San Marcos, CA 92069

Arkla Industries, Inc.
Box 534
Evansville, IN 47704

Artcraft Industries
3500 Walnut St.
McKeesport, PA 15132

B

Barnes Engineering
30 Commerce Rd.
Stamford, CT 06904

Baron Industries, Inc.
4204 N. Brown Ave.
Scottsdale, AZ 95231

Bennett-Ireland, Inc.
Norwich, NY 13815

Borg-Warner Plumbing Products
201 E. 5th St.
Mansfield, OH 44901

Bostitch
Briggs Dr.
East Greenwich, RI 02818

Broan Mfg. Co.
Hartford, WI 53027

Bryant Air Conditioning
7310 W. Morris St.
Indianapolis, IN 46231

C

Calcinator Corp.
1641 Water St.
Bay City, MI 48706

Calmac Mfg. Corp.
Box 710
150 S. Van Brunt St.
Englewood, NJ 07631

Carol Cable Co.
Div. Avnet Co.
249 Roosevelt Ave.
Pawtucket, RI 02862

Carrier Air Conditioning Co.
Carrier Parkway
Syracuse, NY 13201

CDC Chemical Corp.
360 W. 11th St.
New York, NY 10014

CertainTeed Corp.
750 E. Swedesford Rd.
Valley Forge, PA 19482

Champion Spark Plug Co.
P.O. Box 910
Toledo, OH 43601

Chronomite Labs
21011 S. Figueroa St.
Carson, CA 90745

Comfort-Aire
Heat Controller, Inc.
1900 Wellworth Ave.
Jackson, MI 49203

Comfort Enterprises Co.
P.O. Box 323
Leola, PA 17540

Contech, Inc.
7711 Computer Ave.
Minneapolis, MN 55435

Coplanar Corp.
1631 San Pablo Ave.
Oakland, CA 94612

Coughlan Products, Inc.
29 Spring St.
West Orange, NJ 07052

Crest Electronics, Inc.
2634 La Cienega Ave.
Los Angeles, CA 90034

D

DAP, Inc.
P.O. Box 277
Dayton, OH 45401

Dare Pafco Inc.
860 Betterly Rd.
Battle Creek, MI 49016

Darworth, Inc.
Avon, CT 06001

Dasco Products, Inc.
2215 Kishwaukee St.
Rockford, IL 61101

Deeprock Mfg. Co.
P.O. Box 870
Lafayette Parkway
Opelika, AL 36801

W.J. Dennis & Co.
1111 Davis Rd.
Elgin, IL 60120

Dornback Furnace & Foundry Co.
33220 Lakeland Blvd.
Eastlake, OH 44094

Dow Corning Corp.
Midland, MI 48640

Drew International
2019 Brooks
Houston, TX 77026

Duo Therm Div.
La Grange, IN 46761

Dutch's Enterprises, Inc.
4265 S. Pine Ave.
Milwaukee, WI 53207

E

Elmer's
Borden Chemical
180 E. Broad St.
Columbus, OH 43215

Emerson Electric Co.
White-Rogers Div.
9797 Reavis Rd.
St. Louis, MO 63123

E Z Plumb
Div. United States Brass Corp.
P.O. Box 37
Plano, TX 75074

F

Fafco
138 Jefferson Dr.
Menlo Park, CA 94025

Fedders Corp.
Edison, NJ 08817

Filko Automotive Products Div.
F&B Mfg. Co.
5480 N. Northwest Hwy.
Chicago, IL 60630

Filter Dynamics International
18451 Euclid Ave.
Cleveland, OH 44112

Flushmate
Water Control Products/N.A., Inc.
1100 Owendale
Troy, MI 48084

Fram Corp.
105 Pawtucket Ave.
East Providence, RI 02916

Friedrich Air Conditioning & Refrigeration Co.
4200 N. Pan Am Expressway
Box 1540
San Antonio, TX 78295

Frost-King
Thermwell Products Co., Inc.
150 E. Seventh St.
Paterson, NJ 07524

Fuel-Miser
Flair Mfg. Corp.
600 Old Willets Path
Hauppauge, L.I., NY 11787

Fuel Sentry Corp.
79 Putnam St.
Mt. Vernon, NY 10550

Fulton Corp.
303 8th Ave.
Fulton, IL 61252

G

Gainsborough
Continental Industries, Inc.
201 Clinton Ave.
Plainfield, NJ 07063

Geberit Mfg., Inc.
Michigan City, IN 46360

General Electric Co. (Air Conditioners)
2100 Gardiner Ln.
Louisville, KY 40205

General Electric Co.
1 River Rd.
Schenectady, NY 12345

General Filters, Inc.
43800 Grand River Ave.
Novi, MI 48050

General Products Corp.
150 Ardale
West Haven, CT 06516

General Time Corp.
P.O. Box 338
Davidson, NC 28036

Gold Eagle Products Co.
1872 N. Clybourn Ave.
Chicago, IL 60614

W.R. Grace & Co.
62 Whittemore Ave.
Cambridge, MA 02140

Granberg Industries, Inc.
200T Garrard Blvd.
Richmond, CA 94804

GX International
2610 N.E. 5th Ave.
Pompano Beach, FL 33064

H

Handyman Electric Co.
P.O. Box 2658
Burlington, NC 27215

Hartwig-Hartoglass, Inc.
P.O. Box 282 Cairns Ct.
Woodstock, IL 60098

H-C Products Co.
P.O. Box 68
Princeville, IL 61559

Heatilator Fireplace
P.O. Box 409
Mt. Pleasant, IA 52641

Hemco Home Equipment Mfg. Co.
14481 Olive St.
Westminster, CA 92683

Homelite Div. of Textron Inc.
P.O. Box 7047
Charlotte, NC 28217

Honeywell, Inc.
2701 4th Ave. S.
Minneapolis, MN 55408

Hotpoint
2100 Gardiner Ln.
Suite 301
Louisville, KY 40205

Howmet Aluminum Corp.
P.O. Box 40
1617 N. Washington
Magnolia, AR 71753

Humidifier Descaler
Cosco Chemicals Div.
25 Beachway Dr.
Indianapolis, IN 46224

I

In-O-Vent
7295 Cascade Woods Dr.
Grand Rapids, MI 49506

In-Sink-Erator Div.
Emerson Electric Co.
4700 21st St.
Racine, WI 53406

Insto-Impala, Inc.
998 E. Woodbridge
Detroit, MI 48207

Iso Thermics, Inc.
P.O. Box 86
Augusta, NJ 07822

J

Jade Controls
P.O. Box 271
Montclair, CA 91763

Johns-Manville Corp.
Ken-Caryl Ranch
Denver, CO 80217

Johnson Industries
2638 Yates Ave.
Los Angeles, CA 90040

K

Kaiser Aluminum & Chemical Corp.
300 Lakeside Dr.
Oakland, CA 94643

Kem Mfg. Co., Inc.
River Rd. & Maple Ave.
Fair Lawn, NJ 07410

Kool-O-Matic Corp.
1831 Terminal Rd.
Niles, MI 49120

L

Lauderdale-Hamilton
P.O. Box 45
Shannon, MS 38868

Leigh Products
1870 Lee St.
Coppersville, MI 49404

Lennox Industries, Inc.
200 S. 12th Ave.
Marshalltown, IA 50158

Leslie-Locke
Ohio St.
Lodi, OH 44254

Leviton Mfg. Co., Inc.
59-25 Little Neck Parkway
Little Neck, NY 11362

Locke Stove Co.
114 W. 11th St.
Kansas City, MO 64105

M

Macco Adhesives
Glidden Coating & Resins
30400 Lakeland Blvd.
Wickliffe, OH 44092

Madico
64 Industrial Parkway
Woburn, MA 01801

Magic Chef, Inc.
Cleveland, TN 37311

Maid-O-Mist
3217 N. Pulaski Rd.
Chicago, IL 60641

Majestic Co.
P. O. Box 800
Huntington, IN 46750

Malco Products Inc.
361 Fairview Ave.
Barberton, OH 44203

Mameco, Inc.
4475 E. 175th St.
Cleveland, OH 44128

Manville Mfg. Corp.
342 Rockwell Ave.
Pontiac, MI 48053

Mar-Mac Mfg. Co. Inc.
P.O. Box 278
McBee, SC 29101

Maytag Co.
Newton, IA 50208

McCulloch Corp.
5400 Alla Rd.
Los Angeles, CA 90066

Melnor Industries, Inc.
1 Carol Pl.
Moonachie, NJ 07074

Metropolitan Watersaving Co., Inc.
5130 McArthur Blvd. N.W.
Washington, DC 20016

Michlin Chemical Corp.
9045 Vincent St.
Detroit, MI 48211

Micro Lambda
Winter Park, FL 32789

Micro-Por, Inc.
10 Innovation Ln.
Colwich, KS 67030

Midget Louver Co.
800 Main St.
Norwalk, CT 06852

Modern Aire Sun Control Products Inc.
431 4th Ave., S. E.
Rochester, NY 55901

Modern Maid
Box 1111
Chattanooga, TN 37401

Montgomery Ward
618 W. Chicago Ave.
Chicago, IL 60607

Mortell Co.
550 Hobbie Ave.
Kankakee, IL 60901

Mueller Climatrol Corp.
Edison, NJ 08817

N

Nankee Co., Inc.
Engineers Ln.
Farmingdale, NY 11735

National Cellulose Corp.
12315 Robin Blvd.
Houston, TX 77045

Newell Cos., Inc.
916 S. Arcade Ave.
Freeport, IL 61032

Nova
Ecological Water Products, Inc.
142-146 Spring St.
Newport, RI 02840

NuTone Housing Products
Madison & Red Bank Rds.
Cincinnati, OH 45227

O

Owens-Corning Fiberglas Corp.
P. O. Box 901
Toledo, OH 43601

P

Panel Era Mfg. Ltd.
8001 Carpenter Freeway
Dallas, TX 75247

Pease Co.
Ever-Strait Div.
7100 Dixie Hwy.
Fairfield, OH 45023

J.C. Penney Co.
1301 Ave. of the Americas
New York, NY 10019

Phifer Wire Products, Inc.
P. O. Box 1700
Tuscaloosa, AL 35401

Plaskolite, Inc.
1770 Joyce Ave.
Columbus, OH 43216

PPG Industries
1 Gateway Ctr.
Pittsburgh, PA 15222

Presto
National Presto Industries, Inc.
Eau Claire, WI 54701

Preway, Inc.
1430 2nd St. N.
Wisconsin Rapids, WI 54494

Price Pfister
13500 Paxton St.
Pacoima, CA 91331

R

Rapperswill Corp.
305 E. 40th St.
New York, NY 10016

RCR Ltd.
2295 Metropole
Longueuil, Quebec, Canada

Red Devil, Inc.
2400 Vauxhall Rd.
Union, NJ 07083

Reynolds Metal Co.
6601 W. Broad St.
Richmond, VA 23261

Rheem Water Heater Div.
City Investing Co.
7600 S. Kedzie Ave.
Chicago, IL 60652

Rho Sigma, Inc.
15150 Raymer St.
Van Nuys, CA 91405

Ristance Products, Inc.
Industrial Park
Argus, IN 46501

Rockwool Industries, Inc.
3600 S. Yosemite St.
Denver, CO 80237

W. J. Ruscoe Co.
475-485 Kenmore Blvd.
Akron, OH 44301

R-V Lite
Arvey Corp.
3500 N. Kimball Ave.
Chicago, IL 60618

S

S-B Mfg. Co., Ltd.
11320 Watertown Plank Rd.
Milwaukee, WI 53226

Schlegel Corp.
1555 Jefferson Rd.
P. O. Box 197
Rochester, NY 14601

Sears, Roebuck & Co.
Sears Tower
Chicago, IL 60684

Season-All Industries, Inc.
Route 119 S.
Indiana, PA 15701

Siloo, Inc.
393 7th Ave.
New York, NY 10001

A. O. Smith Corp.
Consumer Products Div.
Kankakee, IL 60901

Solar Div.
Champion Home Builders Co.
5573 E. North St.
Dryden, MI 48428

Solar Shelter
P.O. Box 36
Reading, PA 19603

Solar Works, Inc.
4817 1st Ave. S.
Birmingham, AL 35222

Solar-X Corp.
25 Needham St.
Newton, MA 02161

Solite
City of Industry, CA 91748

Speed Queen Div.
McGraw Edison Co.
Ripon, WI 54971

Standard Motor Products, Inc.
37-18 Northern Blvd.
Long Island City, NY 11101

Stant Mfg. Co., Inc.
1620 Columbia Ave.
Connersville, IN 47331

State Industries
Box 307
Ashland, TN 37015

J. P. Stevens & Co., Inc.
P. O. Box 1138
Walterboro, SC 29488

Stiles Corp.
500 Mavis St.
Irving, TX 75060

Sun Electric Corp.
3011 E. Route 176
Crystal Lake, IL 60014

Swingline, Inc.
32-00 Skillman Ave.
Long Island City, NY 11101

Swiss Blinds Div.
Richard Goder Assoc., Inc.
7064 Lyndon
Rosemont, IL 60018

T

Tailored Tuff of Fort Worth
3615 A Camp Bowie Blvd.
Fort Worth, TX 76117

T & S Brass & Bronze Works
128 Magnolia Ave.
Westbury, L.I., NY 11590

Tappan Co.
Janitrol Div.
206 Woodford Ave.
Elyria, OH 44035

TechnoSci Inc.
Box R
Fairfield, PA 17320

Teledyne Mono-Thane
1460 Industrial Parkway
Akron, OH 44310

Thermograte Enterprises, Inc.
51 Iona Ln.
St. Paul, MN 55117

Thermo-mist
Steinen of Carolina, Inc.
(Div. of William Steinen Mfg. Co.)
Route 7, Box 90, Airport Rd.
Kinston, NC 28501

Thermotrol Corp.
29400 Stephenson Hwy.
Madison Heights, MI 48071

Thermtron Products, Inc.
P.O. Box 9136
Baer Field
Ft. Wayne, IN 46809

Thexton Mfg. Co.
7685 Parklawn Ave.
P. O. Box 35008
Minneapolis, MN 55435

3M Co.
Industrial Tape Div.
3M Center
Saint Paul, MN 55101

Trol-A-Temp
Div. Trolex Corp.
725 Federal Ave.
Kenilworth, NJ 07033

True Value Hardware Stores
See your local listing.

Twinoak Products, Inc.
RR 2, Box 56
Plano, IL 60545

U

United States Gypsum Co.
101 S. Wacker Dr.
Chicago, IL 60606

United States Mineral Products
Stanhope, NJ 07874

V

Vimco Corp.
P. O. Box 212
Laurel, VA 23060

W

Warp Brothers
1100 N. Cicero Ave.
Chicago, IL 60651

Warren Group
Div. of Warren Tool Corp.
P. O. Box 68
Hiram, OH 44234

Wartain Lock Co.
20525 E. Nine Mile Rd.
St. Clair Shores, MI 48080

Washington Stove Works
P. O. Box 687
Everett, WA 98201

Wells Mfg. Corp.
2-26 S. Brooke
Fond Du Lac, WI 54935

Western Auto Supply Co.
2107 Grand Ave.
Kansas City, MO 64108

Westinghouse Electric Corp.
Major Appliance Div.
300 Phillippi Rd.
Columbus, OH 43228

Whirlpool Corp.
Administrative Center
Benton Harbor, MI 49022

White-Westinghouse Appliance Co.
930 Duquesne Blvd.
Pittsburgh, PA 15222

Windowseal
Morrell Co.
550 N. Hobbie Ave.
Kankakee, IL 60901

Woodhill Chemical Sales
18731 Cranwood Parkway
Cleveland, OH 44128

Woodings-Verona Tool Works
P. O. Box 126
Verona, PA 15147

Wrap-On Co., Inc.
341 W. Superior St.
Chicago, IL 60610

Wrightway Mfg. Co.
371 E. 116th St.
Chicago, IL 60628

Y

York Industries Corp.
Home Products Div.
6 Hoffman Pl.
Hillside, NJ 07205